Process Quality

Second Edition

Technical Editors

Willie L. Myles
Wharton County Junior College
Wharton, TX

Eugene Theobald
Wharton County Junior College
Wharton, TX

Director, Employability Solutions: Kelly Trakalo
Media Producer: Jose Carchi
Development Editor: Rachel Bedard/Editorial Consultants
Content Producer: Alma Dabral
Director, Digital Studio and Content Production: Brian Hyland
Instructor and Student Supplement Development: Perci LLC dba/Publisher's Resource Center
Product Marketing Manager: Melissa Natali
Manufacturing Buyer: Roy Pickering, LSC Communications
Cover Designer: Carie Keller
Cover Image Credit: Pearson
Editorial and Full-Service Production and Composition Services: Pearson CSC
Editorial Project Manager: SPi Global
Full-Service Project Manager: Gowri Duraiswamy
Printer/Bindery: LSC Communications Owensville
Cover Printer: Phoenix Color

Library of Congress Cataloging-in-Publication Data
Names: North American Process Technology Alliance.
Title: Process quality / North American Process Technology Alliance ;
    technical editors, Willie L. Myles, Eugene Theobald.
Description: Second edition. | [Hoboken, New Jersey]: Pearson
    Education, Inc., [2021] | Includes index.
Identifiers: LCCN 2020036083 (print) | LCCN 2020036084 (ebook) |
    ISBN 9780136424703 (paperback) | ISBN 9780136424796 (epub)
Subjects: LCSH: Process control. | Quality control. | Manufacturing
    processes--Quality control.
Classification: LCC TS156.8 .P7656 2021 (print) | LCC TS156.8 (ebook) |
    DDC 658.5/62--dc23
LC record available at https://lccn.loc.gov/2020036083
LC ebook record available at https://lccn.loc.gov/2020036084

6 2023

www.pearsonhighered.com

ISBN-10:   0-13-642470-8
ISBN-13: 978-0-13-642470-3

Preface

The Process Industries Challenge

In the early 1990s, the process industries recognized that they would face a major staffing shortage because of the large number of "baby boomer" employees who would be retiring. Industry partnered with community colleges, technical colleges, and universities to remedy this situation. These collaborators in education and industry recognized that pretraining for process technicians would benefit industry by reducing the costs associated with training and traditional hiring methods. They recognized that teachers needed consistent curriculum content and exit competencies in order to produce process technology graduates who would be knowledgeable, competent, and able to take over the demands of the field. This was how the NAPTA series on Process Technology was born.

To achieve consistency of exit competencies among graduates from different schools and regions, the North American Process Technology Alliance identified a core technical curriculum for the Associate Degree in Process Technology. This core consists of eight technical courses and is taught in institutions throughout North America.

Instructors who teach the process technology core curriculum, and who are recognized in industry for their years of experience and depth of subject matter expertise, requested that a textbook be developed to match the standardized curriculum. Reviewers from a broad range of process industries and educational institutions participated in the production of these materials so that the presentation of content would address the widest audience possible. This textbook is intended to provide a common national standard reference for the *Process Quality* course in the Process Technology degree program.

The textbook is intended for use in high schools, community colleges, technical colleges, universities, and corporate settings and by anyone desiring an understanding of basic quality concepts and practices. Current and future process technicians will use the information within this textbook as their foundation for work in the process industries. This knowledge will make them better prepared to meet the ever-changing roles and responsibilities within their specific process industry.

What's New!

The second edition has been thoroughly updated and revised.

- **New Content Organization** to enhance the learning process. Content flows from basic Quality concepts (Part 1) to major Quality strategies (Part 2) to Quality tools (Part 3) to application of control charts and everyday process industry needs (Part 4).

- **New Chapter Objectives aligned** with NAPTA core objectives, with links from objective to text page provided.
- **New Chapter** on data collection and representative sampling.
- **New Economics content** to provide basic information that process technicians will need.
- **New Teams content** discussing strategies and "soft skills" that will enhance the ability of process technicians to excel in the workplace.
- **All New Full Color Art Program** with more than 190 fully revised drawings and photos, including all new visuals.
- **Updated coverage** of ISO standards.
- **New Key Term definitions** at the beginning of each chapter and on the text pages where content appears.
- **Updated Review and New Answers Appendix!** Checking Your Knowledge questions have been updated to meet new chapter objectives and now answers are included in an Appendix.
- **ALL NEW Instructor Resource Package** including lesson plans, test banks, review questions, PowerPoints, a correlation guide to NAPTA curriculum, and a Transition Guide from first edition to second edition.
- **ALL NEW Pearson eText** available to facilitate digital learning!

Organization of the Textbook

Each chapter has the same organization.

- **Objectives** for each chapter are aligned with the revised NAPTA curriculum and can cover one or more sessions in a course.
- **Key Terms** list important words or phrases and their definitions, which students should know and understand before proceeding to the next chapter.
- The **Introduction** provides a simple introductory paragraph or introduces concepts necessary to the development of the chapter's content.
- Any of the **Key Topics** can have several subtopics. Topics and subtopics address the objectives stated at the beginning of each chapter.
- The **Summary** is a restatement of important points in the chapter.
- **Checking Your Knowledge** questions are designed to help students do self-testing on essential information in the chapter.

- **Student Activities** provide opportunities for individual students or small groups to apply some of the knowledge they have gained from the chapter. These activities generally should be performed with instructor involvement.

Acknowledgments

A huge thank you goes to Willie L. Myles and Eugene Theobald, the Technical Editors for *Process Quality*, 2nd edition. Their expertise as Process Technology educators, coupled with their industry experience from the Dow Chemical North America Company and the Shell Oil Company, respectively, allowed them to guide this book steadfastly to completion. We are grateful for their determination, focus, and sense of humor as they met deadlines in order to get this book updated and published for use.

The second edition of this series was made possible with the support of the entire NAPTA Board of Directors and, in particular, the leadership and dedication of Executive Director Eric Newby. We continue to owe a debt of gratitude to Martha McKinley, interim co-chair of the NAPTA Curriculum Committee, who has dedicated untold hours to the pursuit of excellence for the second edition of the NAPTA series.

Thanks go out to Jim Lamar, the primary contributor to the first edition of this book, for the groundwork he laid. His understanding of Quality Management and his enthusiasm for training others are still evident in this revised edition.

We also wish to acknowledge the following organizations and the dedicated personnel who participated in the production of this textbook. Their contributions toward making this a successful project are greatly appreciated.

Process Quality Contributors and Reviewers

Jim Lamar, primary contributor, *Process Quality* 1st edition; Saint-Gobain NorPro, Texas

Josephine A. Allen, Ph.D., St. James Parish Public School System Career and Technology Center, Louisiana

Allen Baragar, Brazosport College, Texas

BASF North America (Art Program)

Curtis Briggs, INEOS, Texas

Nat Byrom, Valero Energy Corporation, Beaumont/Port Arthur, Texas (Retired)

Kathy Brossette, Baton Rouge Community College, Louisiana

Melissa D. Lassiter Collins, BP Learning & Development, Texas

Regina Davis, The Dow Chemical Company, Freeport, Texas (Retired)

Jerry Duncan, College of the Mainland, Texas

Raymond Fisher, Fisher Consulting, Inc., Baton Rouge Community College, Louisiana

Cleve J. Fontenot, BASF Corporation—The Chemical Company, Louisiana

Henry Haney, Kenai Peninsula College, Alaska

Rhea Jones, ConocoPhillips, Oklahoma

Karen Kupsa, College of the Mainland, Texas

Carol Lamar, Independent Reviewer, Texas

Anthony Pringle, Remington College, Texas

Dennis Rygaard, The Dow Chemical Company, Texas (Retired)

Shell International Limited (Art Program)

Lawrence Wick, Victoria College, Texas

This material is based upon work supported, in part, by the National Science Foundation under Grant No. DUE 0532652. Any opinions, findings, and conclusions or recommendations expressed in this material are those of the author(s) and do not necessarily reflect the views of the National Science Foundation.

Contents

v

Part 4 Variability, Control Charts, and Process Capability

Chapter 1
Introduction to Process Quality

"Quality is neither mind nor matter, but a certain entity independent of the other two . . . even though Quality cannot be defined, you know what it is."

~ ROBERT MAYNARD PIRSIG

Objectives

Upon completion of this chapter, you will be able to:

1.1 Identify characteristics that help define quality. (NAPTA Quality, Overview 1*) p. 2

1.2 Describe the history of the quality movement in the United States. (NAPTA Quality, Overview 2) p. 4

1.3 Compare and contrast the different philosophies of Deming, Juran, Crosby, and others. (NAPTA Quality, Overview 3, 4, 5, 6) p. 5

1.4 Identify key quality concepts employed in the process industries today. (NAPTA Quality, Overview 1) p. 11

Key Terms

Control chart—graphical presentation of statistical data that can be used to follow and identify process variations and problems, **p. 4**

Process Safety Management (PSM)—an OSHA standard, 29CFR1910.119, concerning process safety management of highly hazardous chemicals, **p. 2**

Quality—conformance to all aspects of customer requirements, **p. 2**

Statistical process control (SPC)—statistical procedures that keep track of a process in order to reduce variation and improve quality, **p. 11**

*North American Process Technology Alliance (NAPTA) developed curriculum to ensure that Process Technology courses will produce knowledgeable graduates to become entry-level employees in process technology. Objectives from that curriculum are named here in abbreviated form. For example, "(NAPTA Quality, Overview 1) means that this chapter's objective 1 relates to objective 1 of the NAPTA curriculum overview on quality.

Statistical quality control (SQC)—the application of statistical techniques to the output (quality) of a process, **p. 11**

Total quality management (TQM)—the pulling together of many different components of quality into a single program (sometimes referred to as TQC—total quality control), **p. 11**

1.1 Introduction

Quality conformance to all aspects of customer requirements.

Quality is a key component of the success of any business today. Starting up a new business is difficult. Staying in business for the long haul is yet another story. Being the low-cost producer may give your business an edge in capturing a sale, but getting the customer to come back to you for the second and third sale will require that, in addition to a good price, you provide a quality product as well as quality services. The purpose of this textbook is to present you with an overview of the various components of quality and prepare you for your role in supporting quality in the process industries.

Process Safety Management (PSM) an OSHA standard, 29CFR1910.119, concerning process safety management of highly hazardous chemicals.

A good quality program begins with safety. Safety first, then quality. Department of Labor's Occupational Safety and Health Administration (OSHA) **Process Safety Management (PSM)**, designated as OSHA 1910.119, is designed to manage the integrity of operating systems and processes handling hazardous substances by applying good design principles, engineering, and operating practices. Implementation of the 14 elements of PSM forms the foundation for a sound comprehensive safety program, which is crucial to creating a good quality program. It is each process technician's responsibility to work safely and to ensure compliance with both state and federal governmental regulations as they pertain to their operating unit for OSHA 1910.119, Process Safety Management of Highly Hazardous Chemicals.

Defining quality is a daunting task, yet this book attempts at least to describe quality in terms of what it means to those in the process industries. Part of what makes defining quality difficult is that all of the answers you receive may in fact be correct because quality is a broad concept.

Think about quality as a jigsaw puzzle. There are many pieces to the quality puzzle. Each piece fills in a certain gap. If any one piece is missing, then the picture is incomplete and unclear. Only by filling in every single piece can one grasp the total quality picture. For the sake of illustration, the various quality topics in this textbook are represented by six different puzzle pieces:

1. Statistics
2. Customers
3. Management systems
4. Improvement strategies
5. Root cause analysis
6. Quality costs.

As you go through each chapter, check to ensure you understand where each topic fits within the puzzle. This chapter lays the groundwork for the rest of the book by discussing the history of the quality movement, identifying key quality concepts, and promoting understanding of the technical terminology used in the field of quality. Additionally, this chapter lists and briefly describes the kinds of tools covered in more depth throughout this book.

Quality Characteristics

Quality can be characterized in many ways:

- As conforming to requirements
- As characterized by consistency
- As superior performance
- As being better than the competition

- As an absence of defects
- As meeting expectations
- As whatever the customer wants it to be.

Ask a dozen people on the street to define quality, and you will probably receive a dozen different answers, including some of the ones just provided. You can combine many of these concepts to define quality as the performance, products, and services that consistently meet or exceed the expectations of the customer by doing the right job, the right way, the first time, every time (Figure 1.1).

Figure 1.1 Companies are in competition with one another. The company with the best quality products and services will WIN.

CREDIT: Hurst Photo/Shutterstock.

Let us examine some of these concepts further.

- Product quality versus total quality: Quality cannot be characterized as just the product that is purchased—the definition of quality must include the service that came with the product. Most of us have had unpleasant experiences shopping for something. Often, we decide where to shop based on our past service experiences or even based on the service experiences shared with us by friends and family.

- Performance: Quality must be about long-term performance as well as working in the present. Have you ever bought something that worked great on day one but failed within a matter of days or weeks?

- Conformance: Quality must also include conformance to requirements. In the process industries, the customer requirements are typically called product specifications. These product requirements are guaranteed to the customer. Figuring out what the customer's expectations are and translating those expectations into specifications is an important part of a quality program.

- Meet or exceed: Meeting the expectations of the customer may *keep* you in business. Exceeding the expectations of the customer may in fact *help grow* your business. Giving customers more than they ask for and more than is expected is often what it takes to ensure that customers will come back again and will speak positively about the experience.

- Relative quality: Like beauty, quality is in the eye of the beholder. Different customers have different requirements and different expectations.

- Consistency: In the process industries, the products must produce the same results every time. Your customers would no more tolerate inconsistency in their dealings with you than you would when you visit a restaurant. Because consistency is such an important concept when talking about quality, this book will emphasize it throughout.

If you focus on only one aspect of quality, you lose. You may lose one customer. You may lose one sale. You may lose your business. Success requires that you view quality as a multifaceted concept seen from the customer's perspective. Ultimately, you are in business to make money. The quality movement is not just an activity to perform when you have nothing else to do. It is not about creating busywork to make sure you are earning your pay. Quality is one of the key components in making your business successful. A successful business is one that makes money today and into the future.

1.2 The Quality Movement in the United States

Control chart graphical presentation of statistical data that can be used to follow and identify process variations and problems.

In the 1920s, Walter Shewhart (1891–1967) showed his boss at Western Electric a one-page memorandum that pictured what is called today a **control chart**, along with a brief description of how the chart could be used to characterize process variation (Figure 1.2). This document was the beginning of modern-day process quality control.

Figure 1.2 Sample control chart.

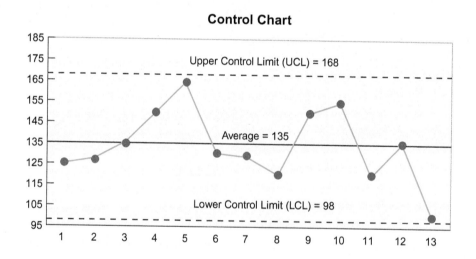

Everything known today about the use of control charts to understand and better control processes is based on Dr. Shewhart's work. His second book, *Statistical Method from the Viewpoint of Quality Control*, published in 1939 by the Graduate School of the Department of Agriculture, is the original work on the subject. As you will see in later chapters, when control charts are mentioned in technical training, they are often referred to as Shewhart control charts, paying homage to their originator.

One of the people who saw and appreciated Dr. Shewhart's work was Dr. W. Edwards Deming, who was working for the U.S. Department of Agriculture in the late 1920s and throughout the 1930s. In fact, Dr. Deming edited Dr. Shewhart's book. Dr. Deming was sent to Japan in the late 1940s to work on the 1951 Japanese census. He accepted an invitation from the Japanese Union of Scientists and Engineers (JUSE) to speak about statistical process control (SPC).

Over the following years, he continued to work with the Japanese as they rebuilt their nation after the destruction of World War II. Through the work of Deming and others, Japan began an intensive, society-wide effort to improve its quality and productivity. The effect of this focused effort was a major economic shift in which Japan's image changed from producer of cheap and breakable products to producer of high-quality affordable products.

Their success was so great that in 1980, NBC aired a white-paper documentary entitled *If Japan Can . . . Why Can't We?* This television program, available on the internet, provided a clear view of how Japanese industries became a powerful force in the world market by their approach to quality. For example, Japan outstripped United States automobile manufacturers in car sales. The program points out several key differences between Japanese and U.S. manufacturing companies at this time:

- Japanese car manufacturers looked for ways to meet the new requirements, while United States car manufacturers resisted environmental regulations and attempted to influence lawmakers to stop regulations.

- Japan was revolutionary in their use of employees in quality improvement. They recognized the importance of not reducing their workforce as quality improvements were gained. This principle helped to engage employees as stakeholders who would benefit from the quality improvement process, rather than be put out of work.

- Japan modernized its 20-year-old equipment while American companies were still running product in 50-year-old plants. When companies like Westinghouse and Motorola did start trying to catch up to the Japanese economic threat, huge investments of capital were required.

- United States manufacturers held back on design and production of electric cars and hybrid vehicles. Hybrid vehicles were largely developed and introduced by the Japanese, and once again U.S. companies followed behind.

When Japan's lead in the global market began increasing, American companies sought the help of experts such as Dr. Deming, Joseph Juran, and Phillip Crosby. As each of these experts promoted his brand of quality improvement as the brand of choice, technical jargon in the quality field grew. Companies often aligned their quality improvement efforts with their chosen teachings, whether of Deming, Juran, Crosby, or others. The following is a brief look at the philosophies of the most recognized quality experts.

1.3 Quality Gurus
Dr. W. Edwards Deming (1900–1993)

The first quality expert is Dr. W. Edwards Deming, whose work in Japan is legendary. The seminars he taught to the Japanese Union of Scientists and Engineers (JUSE) were the beginnings of the Japanese post–World War II rebuilding efforts.

The JUSE board of directors established the Deming Prize to honor the person it credited with having more impact on Japanese manufacturing than any other person not of Japanese heritage. This coveted prize is still one of the highest honors a company can attain. In nearly 70 years, only four American companies have achieved this honor. In Deming's most recognized work, *Out of the Crisis*, he strongly advocated the use of statistical expertise for those seeking to improve the quality of their processes. He went so far as to state that anybody teaching control charts should have at least a master's-degree level of statistical knowledge plus experience working with these charts. Entire chapters are devoted to various statistical topics.

As you will see shortly, this view was not necessarily shared by other quality gurus. He never really characterized quality in a concise manner, instead, he spent nearly an entire chapter talking about how hard it is to define quality and suggesting that quality can be characterized only by the customer. Deming was vocal about his views on management. He advocated the elimination of managing by objectives, focusing on zero defects, and even performance appraisals.

Deming stated that short-term profits were not reliable indicators of management performance and suggested that such short-term focus defeats constancy of purpose and inhibits long-term growth. He even voiced his opinion that tying chief executive officer (CEO) compensation to short-term results would result in CEOs finding it unrewarding to do what was right for the company long term.

Deming was not opposed to the use of advanced technology, but he made it clear that the use of computers, robots, and gadgets would not save American industry. Only hard work and application of his improvement methodologies could accomplish that great feat. He advocated doing things manually in order to get the worker closer to the product. As a proponent of training and continuing education, he was adamant that every process must be constantly and continually improved. He stated that putting out fires is not improvement; instead, it only puts the process back to where it should have been in the first place.

He recognized that improvement would best be accomplished by true teamwork but declared that annual ratings instilled fear and defeated the teamwork. Deming recognized that the involvement of everyone in a company was needed to accomplish the quality transition, furthermore, he advocated breaking down of barriers between the layers of a company. He felt that the barriers were put in place by misguided management who saw employees as the root of the problem.

Deming stated in an interview that management caused 85 percent of a company's problems. In that same interview, he also declared that inspecting a product cannot make it a quality product—you have to make it right in the first place. He saw the divide between management and workers as a key obstacle to success.

During his later years, he was often sought after as a consultant in large American companies. If the CEO would not show up for the meeting, Deming would leave, believing that top-down action was required for success.

His philosophy about quality is best summarized in his famous 14 points:

1. Create constancy of purpose toward improvement of product and service, with the aim to become competitive, stay in business, and provide jobs.
2. Adopt the new philosophy. We are in a new economic age, created by Japan. Western management must awaken to the challenge, must learn responsibilities, and take on leadership for change.
3. Cease dependence on inspection to achieve quality. Eliminate the need for inspection on a mass basis by building quality into the product in the first place.
4. End the practice of awarding business on the basis of price tag. Instead, minimize total cost. Move toward a single supplier for any one item on a long-term relationship of loyalty and trust.
5. Improve constantly and forever the system of production and service, to improve quality and productivity, and thus constantly decrease costs.
6. Institute training on the job.
7. Institute leadership. The aim of leadership should be to help people, machines, and gadgets to do a better job. Supervision of management is in need of overhaul, as well as supervision of production workers.
8. Drive out fear, so everyone may work effectively for the company.
9. Break down barriers between departments. People in research, design, sales, and production must work as a team to foresee problems of production and problems that may be encountered with use of the product or service.
10. Eliminate slogans, exhortations, and targets for the workforce that ask for zero defects and new levels of productivity.

11. **a.** Eliminate work standards (quotas) on the factory floor. Substitute leadership.

    b. Eliminate management by objective. Eliminate management by numbers, numerical goals. Substitute leadership.

12. **a.** Remove barriers that rob the hourly worker of the right to pride of workmanship. The responsibility of supervisors must be changed from stressing sheer numbers to quality.

    b. Remove barriers that rob people in management and engineering of their right to pride of workmanship. This means, *inter alia*, abolishment of the annual merit rating of management by objective.

13. Institute a vigorous program of education and self-improvement.

14. Put everybody in the organization to work to accomplish this transformation. The transformation is everybody's job. (Courtesy of *Out of the Crisis* by W. Edwards Deming, pages 23–24.)

IN A NUTSHELL

Deming
Advocated teamwork, effective leadership, elimination of numerical targets and quotas, and reliance on statistics to drive decisions.

Joseph Juran (1904–2008)

Joseph Juran, who characterized quality as fitness for use, is the second guru. Like Shewhart, Juran worked at the Western Electric Hawthorne Works in the 1920s. The JUSE invited Juran, as they had invited Deming, to travel to Japan and assist in the rebuilding of its economy after World War II. Juran gave a series of lectures there in the 1950s.

His works include *Juran's Quality Control Handbook* (1951), which focused on the statistical and technical aspects of quality improvement, and *Juran on Leadership for Quality* (1989), which was directed at managers in corporate America.

While Deming was a huge proponent of the use of statistics, Juran focused more on managerial issues and the use of breakthrough teams to accomplish quality improvement. Juran was not against the use of statistics, but several times in his book on leadership, he warned that managers should be careful that statistics do not become an end in themselves. Juran recognized, as did Deming, that it is all too easy for managers to manage the numbers instead of managing the process and the people.

Juran was a strong proponent of the use of flowcharts, Pareto diagrams, and process capability indexes by project teams. In his leadership book, and his training videos from the 1980s, he declared that all improvements are accomplished only project by project.

Juran certainly did not invent the Pareto principle, also called the 80-20 rule, but he may have been the first to popularize its use in characterizing problems as either the 20 percent "vital few" issues that cause all the grief or the 80 percent "trivial many" issues that should be looked at, but with a lower priority.

Unlike Deming, who was adamant about eliminating all numerical goals, Juran taught that goals could be acceptable as long as they were legitimate, measurable, attainable, and equitable. He was an advocate for training at all levels and in all functions in a company, suggesting that quality training must be a prerequisite for advancement.

He was especially keen on ensuring that managers should be the first to acquire quality training in order to lead the quality revolution. He taught that quality improvement must be established as a continuing process that goes on year after year and that management's job is to create the proper atmosphere and environment to allow for continuous improvement.

A prime message of the 1989 handbook, *Juran on Leadership for Quality* is that management must take the lead in implementing quality. Management must be the first to be trained in quality, must be the first to serve on quality improvement teams, and must personally and extensively participate in the entire quality improvement effort if there is to be any hope of success. Everyone must participate in the quality effort, from engineers to process technicians to the clerical staff, but managers must lead the quality effort.

Juran suggested that attempting to delegate the quality function to subordinates had been tried many times with disappointing results. He recognized the need for employee involvement and the value that workers on the floor brought to the system but strongly asserted that without the leadership and participation of committed managers, the effort would ultimately be in vain.

He proposed a trilogy of managerial processes for managing quality:

- Quality planning—Who are your customers? What do they need? What product features could the company provide to meet that need? What processes does it need to put in place to achieve those product features? How does it institutionalize these processes in the workplace?

- Quality control—Evaluate actual quality performance. Compare it to desired quality performance. Act on the differences.

- Quality improvement—Organize to achieve quality improvement. Identify problem areas. Establish project teams to address the issues. Provide the resources, training, and motivation needed to achieve success.

People's jobs are sometimes impacted by the planning and control phases of a process, but most often they are called upon to improve an existing process or product. Juran taught quality improvement through the use of breakthrough teams—teams that seek not an incremental improvement in a process but a breakthrough to a new level.

Juran's philosophy regarding breakthrough teams was that they are quality practitioners on a series of journeys. During the first journey, called the *diagnostic journey*, the team seeks to understand the root cause(s) of the problem by following a flowchart like the one shown in Figure 1.3.

Figure 1.3 Example of a diagnostic journey.

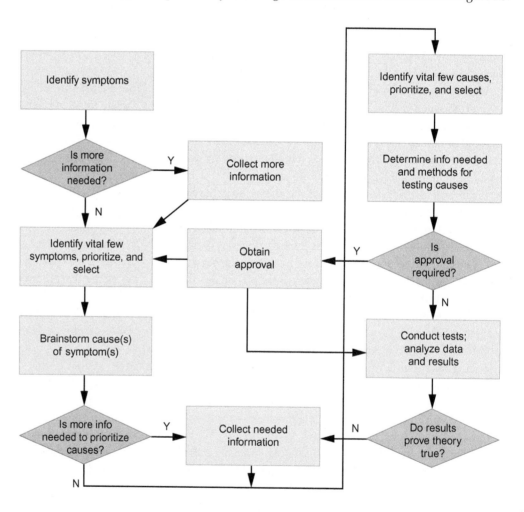

The diagnostic journey begins with gaining an understanding of the symptoms—exactly what is it that management or the customer is upset about? What is the complaint? Next, theories about what might be causing these symptoms are listed, usually by employing techniques such as brainstorming. With a list of possible causes in hand, the team then works on testing its theories. Teams may employ many tools to either validate or invalidate a potential cause.

Once the team has a list of validated root causes, it has completed its diagnostic journey and can move to the *remedial journey*, which is where it remedies the situation. An example flowchart for the remedial journey is shown in Figure 1.4.

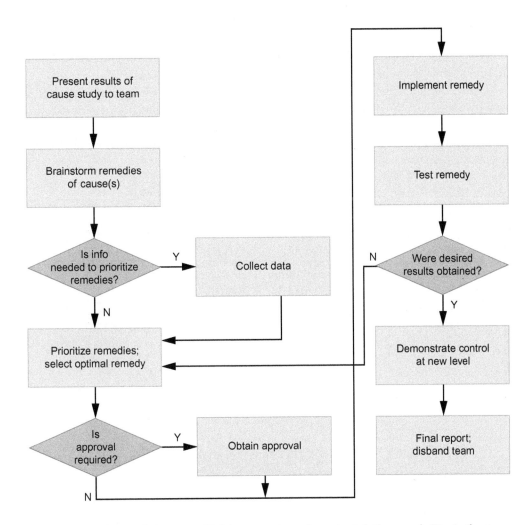

Figure 1.4 Example of a remedial journey.

The basic flow of the remedial journey is rather straightforward: First, the team solicits options for establishing a remedy. It might use brainstorming techniques, internet searches, and/or past history to get started. Then it needs to find a way to evaluate the proposed solutions. Which one costs the most? Which one does a better job of addressing the root cause? Which one can be implemented the most quickly and most easily?

Once the breakthrough team has narrowed down the list of possible solutions and prioritized them, it needs to test them. Given that it is not always possible to test every proposed solution, it may be desirable to run a small-scale test or experiment to validate that a solution actually works. Otherwise, a team might find out that it invested time and money to fix a cause that was not the greatest contributor to the problem after all. Once its solution(s) have been validated, the team must take steps to establish control systems to maintain the gains. Too many times, companies end up fixing the same problems over and over again.

IN A NUTSHELL

Juran

Advocated strong leadership participation and the use of breakthrough teams to identify and eliminate problems.

Philip Crosby (1906–2001)

The third guru in the world of quality is Philip Crosby, who characterized quality as zero defects or, in other words, simple conformance to requirements. Two of his most noted works are *Quality Is Free* (1979) and *Quality without Tears* (1984). You will not find a big discussion about the use of statistics in Crosby's books because his mantra was zero defects, which is accomplished through prevention.

A common acronym attributed to his work is DIRTFT, which stands for "Do It Right the First Time." This was hardly a unique perspective, as Deming and others were saying the same thing. With quality characterized as conforming to requirements, Crosby put a lot of stock in measurement of conformance, focusing on the results of the process.

As was the case with the Deming and Juran, Crosby held that management commitment was absolutely critical to success and that training of the employee base was a key component of achieving that success as quickly as possible. He also recommended the use of project teams as the mechanism to accomplish quality improvement.

Unlike Deming and Juran, Crosby did not offer much in the way of a tangible mechanism to accomplish the quality goals. His books and teachings were more directed toward convincing management that something must be done and helping them organize to achieve quality improvement. His work in bringing the quality crisis to the forefront of management's attention earned him a significant following in industry.

Did You Know?

DIRTFT or "dirty foot" was often displayed on banners and posters and generally brought chuckles, but it got employees' attention and inspired the workforce to follow the quality teaching of Philip Crosby: "Do It Right The First Time." Creating an atmosphere where employees feel free to speak up when they see a problem or have a question is the first step in building a culture of "Do It Right the First Time." The use of the quality tools in this book, combined with the DIRTFT approach, will ultimately improve customer satisfaction, which is essential for staying in business long term.

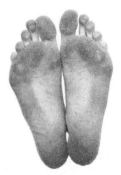

CREDIT: schankz/
Shutterstock.

Crosby suggested a 14-step process to improving quality:

1. Management commitment
2. Quality improvement team
3. Measurement
4. Cost of quality
5. Quality awareness
6. Corrective action
7. Zero defect planning
8. Employee education
9. Zero defect day
10. Goal setting
11. Error-cause removal
12. Recognition
13. Quality councils
14. Repetition (do it again).

IN A NUTSHELL

Crosby

Management drives the organization to achieve zero defects. Meeting the requirements is the only measure of success.

Other Quality Experts

The "big three" quality greats, having firmly established themselves in the quality hall of fame, could arguably be joined by many others who paved the way for quality as we know it today. These other quality greats include the following:

- Peter Scholtes (1938–2009)—author of *The Team Handbook* (1988). In many ways, what Scholtes did was adapt Deming's approach to quality improvement. Scholtes added a structured delivery that made it easier and, to many people, more useful to implement. He also overlaid the Juran concept of project teams to create a holistic approach to quality.

- Tom Peters (1942–)—author of *In Search of Excellence* (1982) as well as many other books. Tom Peters is known as a dynamic speaker and management consultant. His value to the quality transformation in America has not been in adding a new tool to the quality toolbox but in getting everyone excited about quality, from the top of the organization down to the shop floor worker. His PBS specials and videos in the 1980s were widely viewed as a mechanism to infuse energy into organizations. He promotes the concept of a quality champion, stating that at whatever level your champion exists, that is the level of quality success you can expect. He also characterizes quality in terms of the "wow factor," teaching that companies that provide above and beyond what the customer expects are the companies that ultimately succeed.

- Kaoru Ishikawa (1915–1989)—author of *Guide to Quality Control* (1982). His work included implementing basic statistical techniques such as those covered in later chapters of this book. One of these techniques is called an Ishikawa diagram, or fishbone chart, which is discussed in Chapter 11, *Other Basic Quality Tools* .

- Genichi Taguchi (1924–2012)—a Shewhart-Deming disciple. Taguchi is widely recognized as an innovator in the field of designed experiments, and he made many contributions to the field of quality control. He popularized the concept of offline quality control, a term that means you should build quality into the process during the design and construction phase so that you do not experience quality problems during the manufacturing phase. His experimental design methods are often referred to as Taguchi methods, or robust design.

1.4 Key Quality Concepts

The jargon of quality concepts and quality management is not as pervasive in today's industry as it was when these leaders first developed their theories. However, quality concepts are still very much alive, and these concepts are embedded in the framework and planning of major industries. **Total quality management (TQM)** is not a separate topic but rather a melding together of many different quality concepts. Today's students need to understand how TQM works, whether or not they work for a company that actively espouses it. Here is a list of five key concepts related to quality that have stood the test of time and are seen everywhere in industry today:

1. *Management commitment*. If quality does not start at the top, it probably will never get going at all.

2. *Management systems*. Standardization of how things are done to ensure consistency of operation can only serve to drive consistency in the results. ISO 9000 and ISO 14000 are examples of internationally recognized management systems. (Chapter 4, *Quality Management System– International Standard (ISO)*, will provide more information on management systems.)

3. *Statistics*. Most of the works on quality are going to spend a considerable amount of effort showing you how to use statistical tools to run your process as efficiently as possible. Oftentimes, people use the terms **statistical quality control (SQC)** and **statistical process control (SPC)** when discussing the application of statistics to the field of quality.

Total quality management (TQM)   the pulling together of many different components of quality into a single program (sometimes referred to as TQC—total quality control).

Statistical quality control (SQC)   the application of statistical techniques to the output (quality) of a process.

Statistical process control (SPC)   statistical procedures that keep track of a process in order to reduce variation and improve quality.

4. *Teams*. Nearly all the quality greats agree that two heads (or three or four) are better than one.

5. *Training*. If your teams are not properly equipped with the knowledge of management systems, statistics, and improvement methodologies, how can you expect them to succeed?

By studying the concepts and tools of quality, you are preparing yourself for the highest level of company, one that will demonstrate its commitment to quality in general and to your training in the field. The remainder of this book will give you a broad-based understanding of how to apply quality improvement in the process industries. Each of these concepts comes with a wide variety of quality tools to be employed in making improvements. Examples of some of these tools include control charts, process capability, designed experiments, management systems, Pareto charts, and root cause analysis.

All of these tools and many others will be discussed in more detail as you progress through this book. For some of these tools you will need to gain a detailed understanding. For others you will require only a basic understanding. In each case, this text pursues the topic from the viewpoint of how this tool is used by, is impacted by, or affects process technicians.

Summary

Quality is a key component of the success of any business today. This applies to the quality of the product the business sells as well as the service provided during and after the sale. Defining quality requires understanding the needs and expectations of the customer. In the end, quality is making the customers happy so they continue to do business with you. Quality is a multifaceted issue with many pieces similar to a jigsaw puzzle. Each piece plays a role in helping to understand the total quality picture. Leaving out any one piece would result in an incomplete picture.

The quality picture that we have today was developed as a response to the economic threat from the Japanese as they struggled to rebuild their economy after World War II. Many of the quality experts who assisted Japan in their efforts were American scientists and mathematicians such as Dr. W. Edwards Deming and Joseph Juran. The different views of quality worked together to hasten the Japanese economic success. These men and other quality experts such as Philip Crosby and Tom Peters set the foundation for the vast majority of the quality efforts applied in the United States in the 1980s.

The detailed teachings of these experts will fill your quality toolbox with many useful techniques such as control charts, Pareto charts, designed experiments, and capability analyses. Each of these tools has value for quality management and quality improvement efforts. Each can be used to drive continuous improvement. No single concept or tool is sufficient in and of itself, instead, every tool has its own purpose and value to the worker.

To understand how quality is used in the process industries requires a basic understanding of key quality concepts such as statistical thinking, customers, management systems, improvement strategies, root cause analysis, and quality costs. These pieces of the puzzle will be covered in more detail in the following chapters so that by the end of this book you will see the complete quality picture clearly.

Checking Your Knowledge

1. Define the following key terms:
   a. Control charts
   b. Process Safety Management (PSM)
   c. Quality
   d. Statistical process control (SPC)
   e. Statistical quality control (SQC)
   f. Total quality management (TQM)

2. (*True or False*) Quality is an easy term to define.

3. Which of the following concepts must be considered when defining quality?
   a. Performance
   b. Conformance
   c. Consistency
   d. All of the above

4. (*True or False*) A business has to consider the quality of its products and services to ensure long-term success.

5. The "big three" quality greats are:
   a. Juran, Taguchi, and Crosby.
   b. Deming, Juran, and Peters.
   c. Crosby, Deming, and Juran.
   d. Scholtes, Peters, and Taguchi.

6. What are the five key quality concepts?
   a. Teams, training, slogans, meet or exceed, management commitment
   b. Training, teams, statistics, management systems, and management commitment
   c. Management systems, quality councils, repetition statistics, management commitment, and teams
   d. Statistics, management systems, corrective action, training, and teams

7. (*True or False*) The growth of the quality movement in the United States was largely attributable to the success of the Japanese in rebuilding their economy after World War II.

8. Match the following quality tools to their purpose:

Terms	Definitions
I. Control charts	A. Used to judge the acceptability of the process variation against the needs of the customer
II. Process capability	B. Used to standardize the work process so that everybody does the same job the same way every time
III. Pareto charts	C. Used to generate a mathematical model of the process so you can better predict future performance
IV. Designed experiments	D. Used to distinguish between normal process variation (common cause) and problem process variation (special cause)
V. Management systems	E. Used to determine the root cause of a problem prior to initiating any corrective action in order to ensure you fix the right thing
VI. Root cause analysis	F. Used to identify the vital few problems from among the trivial many so you can prioritize where to work first

9. (*True or False*) Dr. Deming was adamant that only by the immediate implementation of computerized robots would the United States ever hope to catch up to the Japanese.

10. What is a diagnostic journey?
    a. An annual conference of quality professionals in Japan
    b. Evaluation of data to determine the root cause of a problem
    c. A review of quality procedures to determine conformance to requirements

11. What is a remedial journey?
    a. The process of selecting the proper remedy for a problem
    b. A brainstorm about a production problem
    c. A set of remedies used at different times

12. (*True or False*) Management commitment is an absolute MUST if the quality effort is to succeed.

13. Zero defects was a key component of the thinking of which quality expert?
    a. Deming
    b. Juran
    c. Crosby
    d. Peters

14. (*True or False*) All three of the quality greats agree that training of personnel was a "nice to have" but not a required part of the quality process.

15. (*True or False*) Management systems are computers used by large companies to control executive and mid-level managers.

NOTE: Answers to Checking Your Knowledge questions are in the Appendix.

Student Activities

1. Write out your own definition of quality. This could be from a personal perspective as a consumer or as a process technician.

2. Compare and contrast the major similarities and differences among the teachings of the "big three" quality greats.

3. Describe briefly the growth of the quality movement in the United States and explain what caused this growth.

4. What is the difference in how to approach statistical quality control (SQC) and statistical process control (SPC)?

5. Explain why teams are an important mechanism for implementing quality.

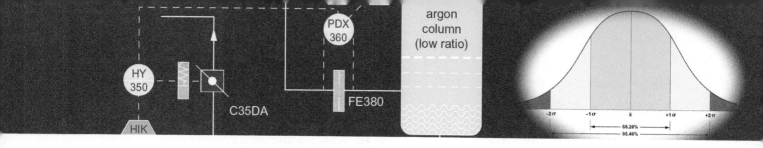

Chapter 2
Total Quality Management and Economics

"Only good things happen when planned; bad things happen on their own."

~ PHILIP CROSBY

 ## Objectives

Upon completion of this chapter you will be able to:

2.1 Explain total quality management and the importance of expressing improvement opportunities in the language of managers—money. (NAPTA Quality, Total Quality Management [TQM] 5*) p. 15

2.2 List and explain the four types of quality costs (internal failure, external failure, appraisal, and prevention), and provide examples of each. (NAPTA Quality, TQM 29–31) p. 17

2.3 Explain what is meant by the terms *price of nonconformance* (PONC) and *hidden factory*. (NAPTA Quality, TQM 6) p. 19

2.4 Describe basic economic concepts and terms that are used in process industries. (NAPTA Quality, TQM and Economics 7–31) p. 20

Key Terms

Appraisal costs—the cost of checking performance to attempt to catch mistakes before they get out the door, **p. 17**

External failure costs—the cost of mistakes that make it out the door to customers, **p. 17**

---

*North American Process Technology Alliance (NAPTA) developed curriculum to ensure that Process Technology courses will produce knowledgeable graduates to become entry-level employees in process technology. Objectives from that curriculum are named here in abbreviated form. For example, "(NAPTA Quality, TQM 5)" means that this chapter's objective 1 relates to objective 5 of the NAPTA curriculum about total quality management.

Hidden factory—the portion of the factory that produces non-value-added product, **p. 20**

Internal failure costs—the cost of mistakes that were caught before they got out the door, **p. 17**

Prevention costs—the costs of improving processes so that they do not have to be checked to catch failures because everything was done right the first time, **p. 17**

Price of nonconformance (PONC)—a term coined by Philip Crosby used to describe the costs of poor performance and failing to meet the customers' needs, also called *cost of poor quality (COPQ)*, **p. 19**

2.1 Introduction

Total quality management is based on the belief that everyone in an organization shares responsibility for the quality of the goods or services sold to the customer. Promote quality, and profits will follow. This theme is evident throughout the quality literature. Unfortunately, many managers have little quality training. Unless the customer complaints result in returned product or lost business, management may not equate quality problems to profits. In some ways, adopting a quality business attitude requires a leap of faith.

Quality experts believe there is value in doing the right thing because it is the right thing to do, and they trust that results will justify their decisions. Sometimes managers have this attitude, while others do not. Over the course of your career, you will probably work for both kinds of manager.

Because the language of management is money, this chapter will give you some basic financial information (Figure 2.1). It will examine four different types of quality costs (internal failure, external failure, appraisal, and prevention) and how these costs affect you as a process technician. The term *price of nonconformance* (PONC) will be defined, and the negative impact of PONC on business will be explained. The concept of the hidden factory will be discussed, and basic economic terms will be explained.

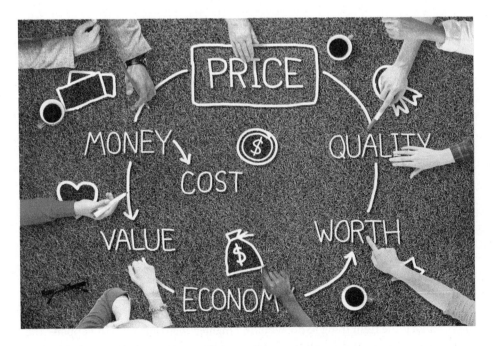

Figure 2.1 The language of management is money.

CREDIT: Rawpixel.com/Shutterstock.

The language of finance can be filled with acronyms and unfamiliar terms: macroeconomics, microeconomics, economies of scale, P/E ratio, return on investment, and so on. You could obtain a 4-year college degree in economics or finance and still have more to learn, but this chapter will focus on costs and economics in relation to quality.

Money: The Language of Management

Money is the language of management. There will be times when you know about an improvement opportunity that will make your process better, but management will not implement the project because there is no guaranteed return on its investment. Sometimes that decision is tough to accept, but employees need to accept that not every improvement opportunity will be implemented.

At least part of a quality effort has to be an honest evaluation of the benefits an improvement might bring. Let us say a process runs at 99 percent purity now. You and your coworkers have an idea that will increase the purity of the product to 99.5 percent. If this idea does not cost any money, management will be pleased to support it. If it does cost money, then management can be expected to question the value of the improvement to the business.

Will this change in the process make the product worth more money to customers, or are they perfectly content with the product as it is now? Will the change in the process result in a reduction of raw materials, reduced run time, or less maintenance? Will it in some other way reduce costs and save the company hard dollars? These questions will be asked whenever a proposal for change is brought forward. Once the benefits are understood and it is known how long it will take to recoup the investment, management will make a decision.

Let us say an improvement idea costs $5 million to implement. If the numbers show management that this project will save the company $10 million next year and every year after that, there is not a manager around who would reject it. On the other hand, if the idea costs only $5,000 but the time to recoup this investment is 10 years or more, management would be far less likely to support the idea.

Companies do not operate with unlimited funds (Figure 2.2A and B). A big part of management's job is prioritizing what to do with the resources at its disposal, and that includes spending the budget wisely. The improvement idea may be great, and from a purely philosophical viewpoint the company should do it because it is the right thing to do. However, if you cannot express the value of the project to managers in their language (money) and show how the idea will benefit the bottom line, the project is not likely to get off the ground.

Figure 2.2 A. Managers must constantly check work results against the bottom line. **B.** Time to payback is a crucial element in getting approval for an improvement idea.

CREDIT: A. sutadimage/Shutterstock.

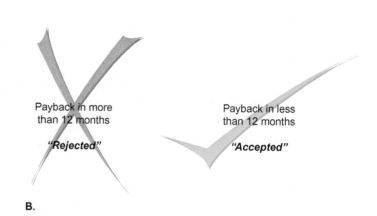

Payback in more than 12 months "Rejected"

Payback in less than 12 months "Accepted"

A.

B.

The hardest part of this concept is the element of strategic importance. If the idea is great, and it is the right thing to do, wouldn't management just make it happen? The answer to that question has to do with the different types of managers that exist. Some managers are visionaries. They will see the long-term benefit and perhaps support the project even if there is no guaranteed return. Many are less visionary and manage "by the numbers." This second type of manager will not support a project if the numbers do not justify it. For them, choices are clearcut, and management decisions are easy.

Your role as a process technician will not likely include approving funding for projects, but it will include constantly working to improve the process. Being able to express your ideas in language that managers understand can help you get your ideas implemented. Knowing how managers think will help you see the improvement opportunities from their perspective. To the extent that the improvements will save money, will make money, or can be accomplished without money, you will make a lot of progress. Be prepared to talk dollars when you talk about improvement opportunities.

2.2 Types of Quality Costs

Let us take a look now at the most traditional of views regarding quality costs. Dr. Juran first categorized quality costs into four distinct "buckets" in his book *Quality Control Handbook*. This organizational structure has been covered in the literature by many of the quality greats through the years.

No company is perfect, and mistakes do happen. Even if you have your laboratory check your product before it goes out, every now and then a bad batch slips through. When a bad batch gets past you, you have to fix the problem with your customers. Knowing that your performance is not perfect, you work on making it better. All of the costs associated with making a quality product (or conversely failing to make a quality product) are referred to as *quality costs*. Quality costs (Figure 2.3) are categorized as follows:

- **Internal failure costs**
- **External failure costs**
- **Appraisal costs**
- **Prevention costs**

> **Internal failure costs**   the cost of mistakes that were caught before they got out the door.
>
> **External failure costs**   the cost of mistakes that make it out the door to customers.
>
> **Appraisal costs**   the cost of checking performance to attempt to catch mistakes before they get out the door.
>
> **Prevention costs**   the costs of improving processes so that they do not have to be checked to catch failures because everything was done right the first time.

When a quality program is first launched, the quality costs might look like the graph shown in Figure 2.3A. In the beginning, you see a high rate of failures, both internal and external. Clearly, something must be done, but what? Costs are already out of control, so how can the company afford an improvement effort? The quality manager tackles the problem by instituting a comprehensive system of checks and balances and double checks and more balances to make sure problems in the plant do not get out to the customer. In just a little while the new quality cost report might look like the one shown in Figure 2.3B.

This shows that the company has succeeded in driving down the external failures by increasing appraisal practices. Unfortunately, overall quality costs have not been significantly reduced. They have just been shifted from external failure to appraisal. The process is moving in the right direction, but it is still not where the company wants to be.

Suppose that a quality manager goes off to training and comes back with a lot of great ideas. With the support of management and full participation from the technical staff and process technicians, the quality manager launches a vigorous quality improvement initiative that includes the following approaches discussed in later chapters of this text:

- Capability studies to understand the process variation and the significance of that variation on the needs of the customer (see Chapter 16, *Process Capability*)
- Control charts throughout the process to help individuals understand when that variation is normal and when to react (see Chapter 13, *Variance and Operating Consistency* and Chapter 14, *Variables Control Charts and Interpretation*)
- Designed experiments to optimize the now statistically stable processes (see Chapter 10, *Group Problem Solving–Designed Experiments*)
- Six Sigma teams to drive breakthrough improvements in the existing processes (see Chapter 8, *Continuous Improvement–Six Sigma*)
- Standardized management systems based on the ISO 9001 standard to create consistency in every aspect of the operation (see Chapter 4, *Quality Management System–International Standard [ISO]*)

- Quality reliability planning studies to ensure that newly developed products and processes are directly linked to the needs of the customer and all quality issues are understood and addressed in advance of implementation (see Chapter 5, *Quality Management–Quality Reliability Planning*)

- Root cause analysis tools to ensure that every problem is solved at the source and prevented from recurring (see Chapter 7, *Continuous Improvements–Root Cause Analysis [RCA] and Corrective Action/Preventive Action [CPA]*)

- Lean tools implemented throughout the organization to ensure that waste is minimized, costs are driven down, and everybody is engaged in the quality transformation (see Chapter 9, *Continuous Improvement–Lean*)

- Quality circles implemented throughout the manufacturing organization to ensure that individuals on the front lines are intimately involved in the quality process.

The quality initiative costs a little bit to implement, but the results show that the money is well spent. The company's prevention costs go up, but the total quality costs are down compared with the baseline. Phase III (shown in Figure 2.3C) is a success.

In fact, after just a few short years of effective implementation, these quality initiatives result in reduced failure rates that can lead to cutbacks on appraisals altogether. By figuring out how to make it right the first time, the company does not have to spend so much time analyzing it after the fact. Staff know where to run the processes and they know the expected results from running the process at optimum levels. Tools are in place to signal when the process is behaving in an abnormal fashion so that problems can be fixed when they occur. As a result, overall quality costs continue to decrease (shown in Figure 2.3D).

Figure 2.3 **A**. Quality costs—initial view, or phase 1. **B**. Quality costs—phase II. **C**. Quality costs—phase III. **D**. Quality costs—phase IV.

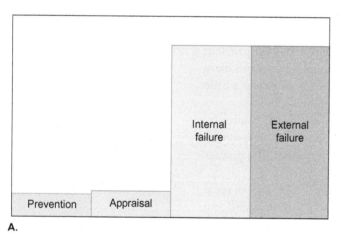

A.

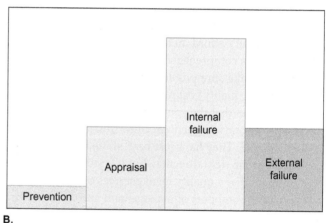

B.

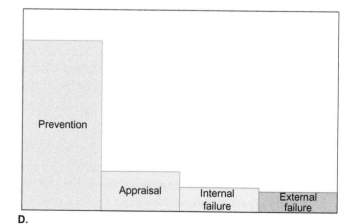

C.

D.

The types of quality costs for each category are named here:

Prevention costs:

- Quality planning
- Capability studies
- Quality improvement projects
- Quality training

Appraisal costs:

- Raw material testing
- Final product inspection and testing
- Audits
- Equipment calibration

Internal failure costs:

- Cost of scrapping the product
- Cost of reworking the product
- Cost of retesting the reworked product
- Delta cost of selling inferior grade product

External failure costs:

- Customer claims
- Product returns
- Product recalls
- Costs of investigating customer complaints
- Loss of business

2.3 Price of Nonconformance

If you were to add up all the various costs of both internal and external failures, you would be able to calculate the total cost of not doing things right the first time. Together these costs are referred to as the **price of nonconformance (PONC)**, a term coined by Philip Crosby. Depending on where you work and which quality expert your particular business unit follows, you may also hear this referred to as the *cost of poor quality (COPQ)*.

In addition to the obvious costs included in this calculation, there may also be less tangible but no less real costs to the business in terms of lost customers, reduced customer trust, and so on. Customers often like to do business with vendors who are ISO 9001 certified (see Chapter 4, *Quality Management System–International Standard [ISO]*). The term "ISO 9001 certified" implies that a company's quality management system will ensure that the vendor is able to consistently make a product that will meet or exceed customer expectations. Companies that are not ISO 9001 certified are more likely to lose business and to fail.

It is difficult to calculate just how much damage can be done by sending your customer a bad batch of product. The cost of replacing the product is easy to calculate. The claim paid to a customer because your product shut down its system or produced bad product in its process is easy to define. However, the costs to the business relationship and your reputation are much more difficult to quantify.

If customers are happy because a business meets and exceeds their expectations, they will have no reason to check out the competition. They will form loyalty to the company because they know they can trust what they get from it. The ultimate price of nonconformance is a business that loses money, trust, and customers. Such a company cannot stay in business for long.

Price of nonconformance (PONC) a term coined by Philip Crosby used to describe the costs of poor performance and failing to meet the customers' needs, also called *cost of poor quality (COPQ)*.

The Hidden Factory

Internal failure costs are just as bad in the long run as external failure costs. The preceding section covered external failures, the real and obvious price paid when bad product is shipped to customers. Internal failures also result in hard dollar expenses, as well as intangible business consequences. What about the costs of internal failures (Figure 2.4)?

Figure 2.4 The cost of internal failures.

CREDIT: Atsushi Hirao/Shutterstock.

Every minute spent making bad product is a wasted minute. Every minute spent blending, reworking, retesting, and trying to deal with a bad product is another wasted minute. The cost of utilities needed to reprocess a bad product is wasted. The cost of the lab materials used to retest the bad product is all wasted. In fact, even the time spent tracking bad product is wasted time. From a customer's perspective, all these wasted resources and all this wasted time add up to a more expensive product. In contrast, if a product is made correctly the first time, all of these expenses can be avoided.

It is useful to think of industry processes as though you have two different factories running, one within the other. There is one factory that makes good-quality, saleable product. There is another, hopefully smaller, factory that makes junk. It can be called the **hidden factory**. It costs money to keep the hidden factory running. The money that a factory diverts to cover the cost of its hidden factory is money stolen from profits.

In some manufacturing plants, hidden factory costs amount to 20 percent of the cost of sales. If the company can reduce costs by reducing waste and reducing nonconforming product, it can make the same profit with a 20 percent reduction in its pricing structure.

If the market is extremely competitive, this difference can mean beating the competition. If a company has great market position and offers an advantaged product, it might be able to keep prices up and pocket the additional profits. Either way, the company wins, and if the company wins, you get to keep your job, so ultimately employees win too.

Hidden factory the portion of the factory that produces non-value-added product.

2.4 Basic Economics

To understand how quality and economics are linked, it is important to understand certain basic economic terms and concepts. Economics is the production, distribution, and consumption of goods and services.

Economics has practical value in business. An understanding of the overall operation of the economic system enables business executives to formulate better policies. An understanding of economic ideas helps the process technician understand the company's goals and support its processes (Figure 2.5). This knowledge also gives individuals, as consumers and as workers, insights on how to make wiser buying and employment decisions.

Figure 2.5 Learning about the overall operation of a business.

CREDIT: Photographic Services, Shell International Limited.

The following economics terms are often mentioned in industries as they pertain to the cost of operating and maintaining a plant (Table 2.1). Process technicians should know the following terms to better understand operating costs of their plant.

Table 2.1 Economics Terminology

Term	Definition
Economics	Production, distribution, and consumption of goods and services
Competition	A market where there are so many sellers that each one acts as though he or she can exert no control over price
Supply and demand	The balance between available product and the market's desire for it; this balance generally determines the asking price
Inflation	A rise in the general level of prices within an economy
Risk	The possibility of loss of value
Standard of living	A minimum of necessities, comforts, or luxuries held essential to maintaining a person or group in customary or proper status or circumstances
Productivity	A demonstrated reservoir of skills, talents, morale, and initiative
Downsizing	Fewer workers are required to produce the same amount output
Profit	The excess of the selling price of goods over their cost to produce
Loss	The amount by which the cost of producing an article or service exceeds the selling price
Revenue	The gross income returned by an investment
Five factors of production	Land, labor, capital, entrepreneurship, knowledge
Assets	An item of value owned
Liabilities	An item which is owed
Accounts payable	Payment due to a supplier for goods received or services rendered
Accounts receivable	Payment due from a customer for goods received or services rendered
Income	The amount of gain over a period of time
Fixed costs	Expenses that stay constant (salaries, wages, benefits)
Variable costs	Expenses that vary (raw materials, utilities)
Operating costs	Fixed and variable costs combined
Gross profit	Profit before operating costs are deducted
Net profit	Profit after operating costs are deducted
Depreciation	The lowering of the price or value of an item

Revenue is the total income produced (Figure 2.6). *Gross profit* is the total revenue after subtracting the cost of producing the products that were sold. An economic *profit or loss* is the difference between the gross profit and the costs of production. *Income* refers to money received, usually on a regular basis, from labor or investments.

Figure 2.6 Revenue is the total income produced.

CREDIT: RomanR/Shutterstock.

Competition is a term that is familiar outside of economics, and its meaning does not change. Competition exists when two or more groups or individuals strive for the same goal and the winning outcome is not shared. In terms of competition, a company that makes a high-quality product will have a competitive advantage over a company that makes a poor one.

Assets are things that are an economic benefit. Cash, inventory, accounts receivable, real estate, and equipment are assets. *Liabilities* are obligations; they could be money that must be paid (accounts payable) or services that must be performed. Related to assets and liabilities is the concept of *depreciation*, which refers to the decreased value of an asset over time.

Operating expenses are those that are required for a company to run. These will include labor, machinery, energy, delivery, and other costs of production. *Fixed costs* are those that do not relate directly to the amount of output. They remain relatively constant (such as rent and machinery). *Variable costs* (such as wages, utilities, and raw materials) change with the company's output.

Some of the costs of operation include the natural resources that the process industry uses. Some industries, like the oil and petroleum industry, exist to process natural resources themselves, but all use natural resources of some sort for running machinery, cooling and heating, and even for documentation. Besides the cost associated with the resources themselves, there may be regulatory costs as well

Inflation refers to the general increase in the cost of goods in an economy. Usually, inflation is expressed as a percentage. The rate of inflation affects the cost of raw materials and labor, as well as the selling price of products. As inflation continues, the cost of living for employees will also rise. The *standard of living* is the relative number of necessities, comforts, or luxuries a person can afford.

Economic *risk* refers to chances that variables may affect an investment or a company's prospects. These variables include exchange rate fluctuations, a shift in government policy or regulations, political instability, or economic sanctions.

The difference between the revenue from sale of output and the costs of all inputs used are either *profit* or *loss*, depending on whether the company has earned more than it has spent or spent more than it has earned.

Accounts receivable are the amounts owed to a company by its customers. *Accounts payable* are the amounts that a company owes to its suppliers. The amounts of accounts receivable and payable are routinely compared as part of a liquidity analysis, to see if there are enough funds coming in from receivables to pay for the outstanding payables.

Most operating plants consist of *integrated units* in order to produce a product that meets the customer quality and deliver expectations. (Integrated processes combine more than one specific unit process into a piece of equipment or a group of work stations operated under unified control.) A problem in any one of the integrated units can have a negative impact on customer satisfaction.

Supply and demand is an important concept in economics. The relationship between the two is defined as the amount of goods and services that are available for people to buy compared to the amount of goods and services that people want to buy. The law of supply and demand says that if less of product than the public wants is available, then companies can charge more for it. Conversely, if the demand for the product is less than what is being produced, the cost per unit will go down.

The relationship between *productivity* and *profitability* can be a simple one. The more productive a company, the more profitable it should be. However, this concept is more complex. Employee engagement may actually be as important to long term profits as short term productivity. Cut-throat, high pressure work environments may initially increase profits, but they ultimately harm productivity. Employees are stakeholders. Employees who feel they matter to the company are more likely to do their best work. If they feel the company is looking for a way to get rid of them (e.g., in a round of downsizing), they will be distracted and discouraged. Employees who are engaged in their work are absent less, have fewer accidents, and make fewer errors.

In order to make a profit, a company needs certain economic inputs. The economic inputs used to make a profit are called factors of production. According to traditional economic theory, there are five main factors of production: land, labor, capital, entrepreneurship, and knowledge.

Summary

As much as it would be wonderful to exhort all businesses to work just on the quality issues and let the profits come, this concept is not practical in the real world. Managers speak their own language, the language of money. If you want them to listen to you, you have to speak their language, because they will not necessarily learn to speak yours. It is important to show management the value to the company of improving quality. To do this you need to be able to describe the value of your projects in terms of payback to the company. Be aware, though, that not every good idea is adopted.

You need to be able to quantify the benefits of reduced failures when you increase your efforts to prevent them. When you can show management the true cost of poor quality, or the price of nonconformance, and show how much potential profit is being wasted in support of the "hidden factory," then you can get management's attention and support.

The factors of production are the resources used in creating and producing goods or services and are the building blocks of an economy. The factors of production are land, labor, capital, entrepreneurship, and knowledge, which are interwoven to create economic growth.

Economics is the production, distribution, and consumption of goods and services. A basic understanding of economics terms is essential to being a well-informed process technician.

Checking Your Knowledge

1. Define the following key terms:
   a. Appraisal costs
   b. External failure costs
   c. Hidden factory
   d. Internal failure costs
   e. Prevention costs
   f. Price of nonconformance (PONC)

2. The language of managers is:
   a. Productivity.
   b. English.
   c. Money.
   d. Greek to most of us.

3. To get quality projects funded, you have to:
   a. Fill out a project authorization form.
   b. Be willing to lead the project.
   c. Convince management that quality is a worthwhile goal.
   d. Describe the value of the project to managers in terms they understand.

4. Which of these costs is not a traditional quality cost?
   a. Cost of capital
   b. Internal failure costs
   c. External failure costs
   d. Appraisal costs
   e. Prevention costs

5. How might one work toward reducing total quality costs?
   a. Implement a standardized quality management system.
   b. Implement Six Sigma.
   c. Implement variability reduction programs (control charting, etc.).
   d. All of the above.

6. Which of these graphs best represents where you want to be in terms of quality costs?

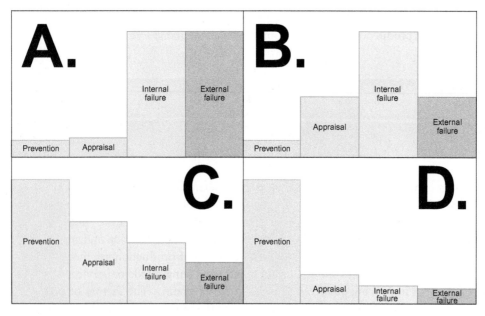

7. Prevention costs include which of the following?
   a. Costs of reworking bad product
   b. Costs of paying customer claims
   c. Costs of quality improvement projects
   d. Costs of testing product

8. The "hidden factory" is:
   a. The portion of your plant that makes poor-quality product.
   b. The part of your plant that you choose not to show customers.
   c. The portion of your plant that is not painted.
   d. The part of your plant that is shielded by a fence from public view.

9. (*True or False*) Companies will invariably implement quality programs based on the philosophical belief that quality is good and then just sit back, expecting the profits to follow.

10. (*True or False*) A basic principle of quality costs is that increasing spending in prevention will ultimately reduce total costs by reducing failure costs.

11. (*True or False*) Quality costs and profits are the responsibility of managers and do not apply to the role of the process technician.

12. (*True or False*) Failure costs are an issue only when the failure occurs at the customer's location.

13. (*True or False*) To work on quality costs effectively requires a degree in economics or finance.

14. (*True or False*) The ultimate price of nonconformance is the inability of a business to stay in business.

15. (*True or False*) Appraisal costs include the costs of conducting quality audits.

16. As an economic term, *productivity* refers to a reservoir of which of the following? (Select all that apply)
    a. Skills
    b. Talent
    c. Morale
    d. Inventory
    e. Initiative

17. (*True or False*) Variable costs include labor, machinery, and delivery.

NOTE: Answers to Checking Your Knowledge questions are in the Appendix.

Student Activities

1. List the costs that contribute to operating expenses.

2. List and explain ways to reduce internal failure as described in the text.

3. List examples of variable and fixed costs.

4. Look online for a documented improvement project for home or business. In a brief report, share the starting point, process, and the ending point with the class.

5. Complete the chart below. Several examples have been given for your understanding.

Personal Example	Concept	Plant Example
Job competition causes you to increase your skills and education to obtain job.	Competition	High competition causes you to be competitive in price and quality or you will go out of business.
High supply of a car brand allows you to purchase the car for a lower price.	Supply	Supply works with demand. Large demand, small supply creates a higher product price.
Super Bowl tickets are in great demand (with limited supply); therefore, they are expensive to purchase.	Demand	High demand/low supply creates high product price. Low demand/high supply creates low price.
	Inflation	
	Risk	

Personal Example	Concept	Plant Example
	Profit	
	Loss	
	Revenue	
	Assets	
	Liabilities	
	Income	
	Accounts payable	
	Accounts receivable	
	Taxes	
	Gross income	
	Net income	
	Fixed costs	
	Variable cost	
	Gross profit	
	Operating expenses	
	Depreciation	
	Natural resources	

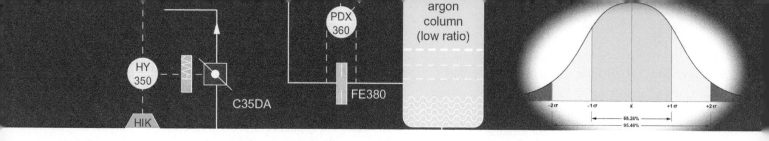

Chapter 3
Customer Service

"Kindness and courtesy are at the root of a positive customer service experience."

~ Shep Hyken

 ## Objectives

Upon completion of this chapter you will be able to:

3.1 Explain the importance of customer service. (NAPTA Quality, Customer Service 1*) p. 27

3.2 Describe the supplier → process → customer (SPC) relationship in terms of a flowchart. (NAPTA Quality, Customer Service 4, 5, 10) p. 29

3.3 Explain the differences between internal and external customers. (NAPTA Quality, Customer Service 2) p. 30

3.4 Compare specifications, requirements, and expectations. (NAPTA Quality, Customer Service 3) p. 31

3.5 Describe an effective customer service interaction. (NAPTA Quality, Customer Service 6, 7, 8, 9) p. 33

Key Terms

Expectations—desires of the customer that might not prevent the product from working as intended but would cause dissatisfaction if not met, **p. 31**

Requirements—the criteria that describe the traits that a product or service must possess in order to function as intended. Performance characteristics of your product, **p. 31**

Specifications—a subset of the requirements that the customer puts into the transaction contract. May also be the translation of the requirement into a measurable output for the purpose of tracking compliance, **p. 31**

*North American Process Technology Alliance (NAPTA) developed curriculum to ensure that Process Technology courses will produce knowledgeable graduates to become entry-level employees in process technology. Objectives from that curriculum are named here in abbreviated form. For example, "(NAPTA Quality, Customer Service 1)" means that this chapter's objective 1 relates to objective 1 of the NAPTA curriculum about customer service.

3.1 Introduction

This piece of the quality puzzle is all about customers. Customers are the reason businesses exist. Without customers, you have no business. For most readers of this textbook, this chapter should be an easy one. After all, every day you play the role of a customer, so you should be customer experts.

Think about the last time you went to the grocery store or the hardware store. For just a few minutes, set product quality aside and look at the service you experienced. Was the parking lot well marked and spacious? Was the store comfortable? How well marked were the aisles and shelves? How clean were the restrooms? Were the employees helpful? How did each person you talked to affect your impression of that establishment? Did some play a greater role than others?

Sometimes customers visit their suppliers. They tour the control rooms and talk to process technicians. Along the way, they are taking mental notes about cleanliness, knowledge, friendliness, and so on. Your behaviors will form a part of their impression of your company (Figure 3.1).

Figure 3.1 Process technicians play an important role in customer visits.

CREDIT: Pressmaster/Shutterstock.

This chapter discusses the importance of service. It describes process in terms of a flowchart in which the company receives inputs from the supplier, adds value by virtue of the process, and then sells outputs to the customer. Our discussion will cover the differences between internal and external customers and outline what distinguishes specifications, requirements, and expectations. Finally, it will describe what makes an effective customer service interaction and show how a process technician plays the role of customer service representative.

The Importance of Customer Service

The age-old question (Figure 3.2) is, "Which came first, the chicken or the egg?" In the world of quality, the question is reframed as, "Which comes first, quality or profits?" Best-in-class companies believe that if you take care of the customer first, then profitability will naturally follow. They also know that the importance of customer service cannot be overstated. If profits are your primary focus, then you might make short-term decisions that increase profits this quarter. However, if you are looking for long-term success, then you will focus on the needs of the customer.

In the quality world, companies are taught that profits are the outcome of doing things right: making the right product and shipping that product in the right container to

Figure 3.2 Questions like, "Which comes first–quality or profits" are often a part of quality discussions.

CREDIT: Hennadii H/Shutterstock.

the right place and with the right paperwork. Companies constantly work to improve their processes to make them the best. If they do these things well, the companies should make money. The primary focus has to be the customer because the customer has the money. The company has to provide a product that convinces the customers to spend their money.

Every contact between you and your customers, every pound of product that you make, every transaction that you conduct should be focused on customer satisfaction. Too often customers are ignored, treated as a nuisance, or treated with such disdain that they choose to look elsewhere. You should never give a customer an excuse to shop elsewhere.

In the quality election, the customer has all the votes. This is not to say that the customer is always right, but the customer is always the customer and has earned the right to be wrong. In all honesty, it is hard to be nice to some customers. Some are just plain difficult. There are customers who complain about a product even though nothing is wrong with it. Sometimes they complain because they think something might be wrong, or they complain in order to negotiate a better price.

If a customer does file a complaint, then the response needs to be a rigorous root cause analysis (described in Chapter 7, *Continuous Improvements–Root Cause Analysis [RCA] and Corrective Action/Preventive Action [CPA]*). If the company is at fault, then it obviously needs to make it right. If the company is not at fault, then there is an opportunity to share the RCA results with the customer and show them the efforts the company is willing to take to improve its processes. Either way, keeping the customer happy must be the goal in order to stay in business.

The response to a customer complaint needs to be thoughtful and respectful and must represent the entire company. For example, imagine a customer issues a complaint concerning incorrect drum labels on the last shipment, and you do not work in that department. It is not appropriate to respond that labelling is not a manufacturing issue, that you do not have any way of dealing with the problem, and that the customer should call the shipping department instead. This kind of response does not typically satisfy the customer. Even though you have helped the customer understand where the problem lies, you have also made the customer do more work in order to get the problem addressed.

The hard fact is that the customer does not care who messed up but just wants it fixed and some assurance that it will not happen again. The correct approach is to show the customer that you care about the problem, assure the customer the problem will be addressed swiftly, and improve the process to prevent future occurrences. The appropriate attitude to adopt when customers voice dissatisfaction is, "Thank you for bringing this improvement opportunity to our attention."

Another response that does not satisfy the customer is, "That's not my job." The attitude is, "I am just a process technician; customer relations is someone else's concern." However, everyone who comes in contact with a customer, directly or indirectly, is a customer service representative.

When it comes to customer satisfaction, how good is good enough? Do you want a 90 percent supplier rating? How about a 95 percent rating? Perhaps a 99 percent rating? Is there such as thing as putting too much effort into this process? Let us do the math. If the electricity worked at your house 99 percent of the time, it would be turned off for approximately 1 day every three months. Would that work for you?

Did You Know?

In 1994, USAir Flight 427, a Boeing 737, crashed while approaching a runway in Pittsburgh. All 132 people on board were killed. The National Transportation Safety Board (NTSB) determined that the cause was a malfunction of the rudder, which moved in the opposite direction from the pilots' commands. After the longest investigation in NTSB history, it was found that the problem with the rudder was the result of a servo valve. This valve usually remains dormant and cold for much of a flight, and it seized after being injected with hot hydraulic fluid. This specific condition had occurred in fewer than 1 percent of the jets that were tested.

A zero defect rate is not just a matter of customer satisfaction. It may save lives!

Nothing short of perfection is the goal when it comes to customer satisfaction. Customer satisfaction is built on more than just product quality. It is built upon the overall experience of doing business. Have you ever been to a department store or grocery store and noticed several employees standing around talking to each other but not to customers? Have you wished they would take the time to find out what you needed and help you find it? When you are the customer, what is important to you?

- Price?
- Product quality?
- Service during and after the sale?
- Ease of transaction?
- Selection?
- Availability of technical information?
- Attitude and personality of the sales force?

We should assume that what is important to us as consumers is also important to our industrial customers.

3.2 Supplier → Process → Customer (SPC)

In this book, the concept of statistical process control, or SPC, is covered in some depth. The concepts of SPC form the foundation for most quality improvement processes. Here is another suggestion of a meaning for this acronym: supplier → process → customer (Figure 3.3).

Whatever job you perform, this is your process. Your process produces an output that goes to a customer. Your process requires inputs that come from a supplier. From the supplier's perspective, you are the customer. From the customer's perspective, you are the supplier. The role you play in this chain of events depends upon your perspective.

Each of these parties needs something in terms of their relationship. Customers have many needs, some of which are communicated verbally and some which are not. They may have needs they do not even know about. They also have expectations that may exceed what they really need. At the least, all customers need the product that they asked for, delivered on time, to the right location, with all of the proper paperwork. In order to fulfill those needs, the people supplying the customer must understand what product is being asked for, when the product is needed, where it is needed, and what kind of paperwork is required.

Figure 3.3 Flow of material from source to product in the pharmaceutical industry.
CREDIT: **A**. Maximiliane/Shutterstock. **B**. Macrovector/Shutterstock. **C**. Mavar/Shutterstock. **D**. LuchschenF/Shutterstock.

A. Raw material (herb foxglove)

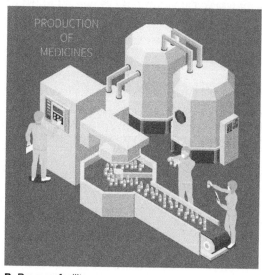

B. Process facility

C. Quality control

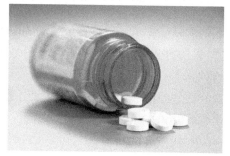

D. Product

Without knowledge of what the customer needs, there is little chance of meeting those needs. Later in this chapter, there will be more about the differences between needs and expectations.

3.3 Internal Customers and External Customers

If the person or organization that receives the output from your process is not part of your company and is paying for the product or service, then this is an external customer. However, many times your customer is part of the same organization, and this person or department is an internal customer. For example, in a plastics plant, the internal customers of personnel working on a turbine that produces steam for production are the workers who use the steam to heat the feedstock that is made into plastic.

Should a worker be just as concerned about pleasing an internal customer as an external customer? Is doing so just as important? Some might answer "No." However, the people or departments within a plant get material from other departments. These "products" become part of the total process of creating top quality product for the outside customer. Having a narrow perspective (silo mentality; Figure 3.4) hurts the company you work for and can damage the end result. Internal relationships are important relationships to maintain. You, your internal suppliers, and your internal customers are all in relationship to serve your external customers.

Figure 3.4 High-level companies teach employees to think of plant working units as customers of each other (e.g., Subprocess B workers are customers of Subprocess A). A silo mentality—the tendency to think only about the needs and profits of one's own working unit—hurts both relationships and products. In the long term, a silo mentality has a negative impact on the overall quality of the product to the external customer.

Beware of Silo Mentality

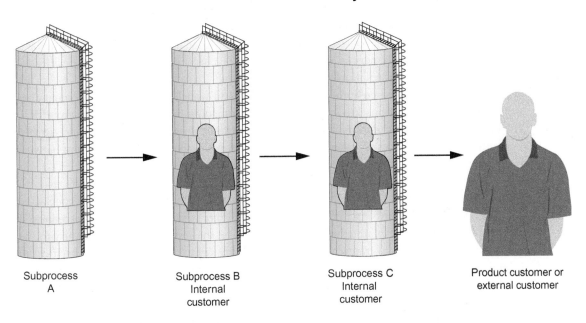

Subprocess A

Subprocess B Internal customer

Subprocess C Internal customer

Product customer or external customer

If a company keeps its internal customers happy but collectively fails to satisfy the external customer, then everyone loses. Sometimes organizations overemphasize the importance of internal customers and in doing so create wonderfully efficient processes for the various units within their companies. They optimize each phase of their operation internally, which is a good concept unless taken too far. It is a mistake to have processes that focus solely on the internal customers and that lose sight of the external customer. This approach may end with the company not providing optimal quality to the external customer. The phenomenon of thinking about only one small part of the overall operation is called *silo mentality*.

An excellent book on this subject is called *Theory Why*, by John Guaspari. The boss in this book solves the "riddle" of quality by getting employees to remember why they do what they do. You do not improve quality by improving productivity in manufacturing or by improving morale in human resources or by improving technology in engineering, or in any other achievement of functional excellence. You improve quality by focusing on adding value to the external customer. Otherwise you make great improvements to each function and still go out of business.

3.4 Specifications, Requirements, and Expectations

Let us revisit three crucial customer service concepts:

- **Requirements**
- **Specifications**
- **Expectations**

Your customer buys a dibutyl product from you. This product is a "filler" in the customer's application and takes up space in its final product mix. The customer has a requirement that this product must flow through a 2-inch (5-cm) unloading hose at a certain rate under ambient conditions in order to unload the truck in a reasonable amount of time. The customer set specifications for the product that can be used to measure its ability to meet

Requirements   the criteria that describe the traits that a product or service must possess in order to function as intended. Performance characteristics of your product.

Specifications   a subset of the requirements that the customer puts into the transaction contract. May also be the translation of the requirement into a measurable output for the purpose of tracking compliance.

Expectations   desires of the customer that might not prevent the product from working as intended but would cause dissatisfaction if not met.

the requirements. In this case, the customer set a viscosity specification, which is a way of measuring how easily a liquid flows. It may be that the customer asked about unloading times, and someone at the company helped with setting the specification. It is often true that customers know what they think they want but are not knowledgeable enough of the technology to understand how to translate those requirements into specifications.

Another chemical, isopropyl alcohol, is used as a cleaning agent. The customer uses this particular chemical to wash away certain organic polymers in its process. The requirement is that the isopropyl alcohol should be an effective cleaning agent, but this is a difficult thing to measure in a lab. Testing shows that your plant's isopropyl alcohol does an excellent job of dissolving organic polymers as long as it is at least 95 percent pure, so a purity specification is set based on this information.

Understanding the customers' requirements and translating those requirements into meaningful specifications are important aspects of doing business. If a company does a poor job of translating requirements into specifications, then customers may end up with product that looks good on paper but does not perform as they expected. Any gap between performance of a product and the customer's expectations is a quality problem.

Expectations are even harder to clarify than requirements are. Customers may not document their expectations. They may not even be able to articulate every expectation. Sometimes expectations are not even known until one is not met. Imagine a scenario in which a customer complains that no one called to let the company know that the product had shipped and was in transit. The customer had never made this contact a specification and had never asked to be notified of shipments made. However, because the customer's other suppliers routinely made this type of phone call, the customer naturally expected that your company would, too. Once an expectation has surfaced, it can be used as an opportunity to work with the customer to clarify requirements and specifications. This can also build customer loyalty by indicating the company's desire to learn and to improve customer service.

Let's take another example. There is a large home improvement store down the road. It carries everything under the sun. Because this nationwide chain buys supplies in railcar quantities, it gets a great volume discount, some of which is passed on to the final customer. Just a couple of miles away is a "mom and pop" hardware store run by John and Jane. With only a single location in a small town, they buy small quantities of the things they believe will sell, so their prices are a little higher. They stay in business because they offer exceptional customer service (Figure 3.5).

Figure 3.5 A. Large volume stores may lack customer service. **B.** Exceptional customer service can create customer loyalty that keeps a small store alive.

CREDIT: A. 06photo/Shutterstock. **B.** michaeljung/Shutterstock.

A.

B.

John and Jane do not just sell drill bits and light switches. They help you understand which drill bit is the best for your application and which light switch will work best for you. If you ask, John will tell you how to wire a three-way switch. They sell a quality service that is not typically found in chain megastores. They sell solutions to household problems. Jane asks what you are working on, not what you want to buy.

Many times, what customers go shopping for is not what they really need. Customers may ask for specific items but may actually need help understanding what would work well for the job.

Did You Know?

The story is told in various management books of the overzealous American purchasing agent who thought he would be tough on his Japanese supplier by setting a failure specification of three parts per 10,000 units. When this company's order arrived from the Japanese supplier, there was a small package in the top of the box with a note that read, "Dear sirs, enclosed please find your first order of 10,000 widgets. We do not know why you desired three bad widgets, but at your instructions we have included them. For your convenience we have packaged them separately so you would not mistake them for good widgets."

Now that is exceeding customer expectations!

Managing expectations is important. Let us say you buy a lightbulb that guarantees it will last for 5 years. You write the date of installation on the bulb when you put it in and find out later that it lasted for only 4½ years. You are not happy because you expected this bulb to last for 5 years. Had you bought the same bulb at the same price but been told that it would last for 4 years, you would have been happy to find out that it lasted for 4½ years. Your expectations would have been exceeded. "Underpromise and overdeliver" is a rule for making customers happy. Exceed customer expectations and they will keep coming back for more.

Expectations change with time. Fifty years ago, a car that lasted 100,000 miles was considered a good car. Today we expect every car to last longer than that.

Times change and so do expectations. By meeting your customer's expectations today, you continually raise the bar on what the customer will expect tomorrow. Consider shopping. It used to be that store hours were set and much more limited than they are today. Most stores were closed on Sundays. Over time, stores began staying open longer and later and most are open seven days a week. With the rise of internet shopping, even those limits have vanished, and consumers can order products any minute of the day or night.

As your customers' expectations change, so must your quality plans and quality programs. Fifty years ago, no one in the process industry was talking about Six Sigma, but it is widely used in industry and business today. What Six Sigma, ISO 9001, and all the other components of a total quality system have in common is their commitment to customer satisfaction. Maintaining high standards and continually looking for ways to improve are ways to build customer loyalty. Companies must accept that change is a natural part of continuing to do business.

3.5 Effective Customer Service Interactions

Most of the time, the discussions of expectations, requirements, and specifications take place outside of the control room and involve the technical staff more than they do the process technician. However, customers come to visit their suppliers. Customers often want to tour and visit the control rooms where the action takes place. Every single person they talk to is a customer service representative for the duration of that visit. Think about a purchasing experience you found to be "top notch." Describe that experience. You might include such things as the following:

- Comfortable temperature
- Well lit areas
- Uncluttered, neat, and tidy environment
- Clean surfaces
- Friendly, helpful staff
- Spacious parking

The next time a customer is coming to visit your plant, create that environment for him or her. Make customers feel special, and they will love coming back. An example of this occurred at a chemical plant back in the early 1990s. After several years of improving their customer visit process and actively managing customer expectations during their visits, customers began commenting about how they looked forward to returning to that plant. Because of this attitude, they began relaxing the scope of their audits and surveys. They came to expect a pleasant visit and excellent performance.

To be sure, you also have to have a good-quality product and a competitive price to stay in business, but managing the less tangible aspects of customer quality plays a key role in convincing customers to keep coming back. Process technicians play a big part in setting the stage for success during these visits. It is good to spend time coaching individuals on proper etiquette during these visits. It is also good practice to clean up the operating areas and then keep them clean between visits. In addition to the physical appearance of a plant, there are other tangible ways to make a positive impression on your visiting customers:

- Be professional
- Be courteous
- Be responsive
- Be polite
- Be friendly, yet businesslike.
- Avoid profanity and inappropriate remarks
- Avoid disparaging remarks about your company
- Avoid negative attitudes.

Remember that you are now the seller. You are being judged, collectively and individually, by the customer. Your goal is to demonstrate to the customer that your company is worthy of the customer's business. If in doubt, apply the golden rule. It works in life, and it works in customer service. Treat your customers as you would like to be treated in the same circumstances.

Your customer service interaction may be direct through a customer visit. It may be indirect by conducting a root cause investigation into a customer complaint. It may be that you are the one putting the paperwork in the appropriate envelope. Whatever your role in the customer contact process, make it a pleasant experience for the customer.

Did You Know?

Sometimes the oldest and simplest rules are the best. When responding to a customer, think how you would want the company representative to speak and act toward you if you were the customer making the call.

"Treat the customers as you would like to be treated"

~The golden rule of customer service

Take a lesson from Disney. This corporation does not have employees at its theme parks; it has cast members. The cast members do not work in an office or at a plant; they are on stage. The personnel department is referred to as "casting." Instead of customers, Disney has guests.

When a customer calls on or comes to visit your plant, you too are on stage. If you need to adopt a stage personality for the visit, then do so. How do actors prepare themselves for a role? They study the role, and they memorize their lines. You should do the same thing. A customer is coming to visit next week. What does this customer buy? What does it make with your product? How is your product packaged and delivered? What is the purpose of the visit? Is the customer auditing you to a company standard or the ISO 9001 standards? What kinds of questions will be asked? What kinds of preparations do you need to make in anticipation of the customer's arrival? Preparing yourself and your location for the customer visit in advance will allow you to relax and host your customer at ease.

The closest thing we have seen to this type of customer attitude in the process industry is a company in the northeast that used to designate one specific carrier to its most critical customers. This carrier would literally have the truck arrive at the customer destination the day before delivery was due, so the driver could have the truck and trailer cleaned. The driver would put on a fresh change of clothes, including a tie, and drive up to the customer's plant an hour before the scheduled delivery time and park on the side of the road. At 10 minutes to scheduled delivery time, the driver would brush his teeth and pull the rig up to the customer's gate, arriving exactly on time, immaculately dressed, and in a truck that was ready for the showroom. There was no doubt in the minds of the customers that the company cared about the impression it was making on their customer.

Summary

All the statistical improvement tools in the world are of little value if a company fails to satisfy the customer along the way. Companies cannot stay in business without customers. Customers have money and that companies need. The company's job is to manufacture a product that will make customers want to buy it.

The statistical tools, quality management systems, and other improvement tools discussed in this book are all focused on improving processes. If the output from these processes does not meet the needs of the customer, then the company will lose in the long run.

The process can be described in terms of a flowchart whereby you receive inputs from your suppliers, add value through your work processes, and then sell this value-added product to your customers. You should also understand the difference between an internal customer and an external customer. Hopefully, you also understand that only by satisfying the external customer can your business succeed.

Covered at some length in this chapter was the difference between requirements (performance characteristics important to a customer) versus specifications (acceptance criteria for your products) versus expectations (things that may not prevent the product from working but will cause customer dissatisfaction if not met). Finally, what makes an effective customer service interaction was discussed, and some practical tips were provided on how to effectively play the role of customer service representative for your company.

Checking Your Knowledge

1. Define the following key terms:
   a. Expectations
   b. Requirements
   c. Specifications

2. Which of the following is important to consider in customer contacts?
   a. Cleanliness of the facility
   b. Availability of technical information
   c. Demeanor of the people
   d. All of these plus more

3. The relationship of your function to the customer can be captured in flowchart form with which acronym?
   a. QFD
   b. SPC
   c. RCA
   d. BOC

4. The recipient of your process output is your:
   a. Accountant.
   b. Supplier.
   c. Customer.
   d. Process.

5. If the person or organization that receives your process output is part of your same company, you call it an:
   a. Internal customer.
   b. Internal supplier.
   c. External customer.
   d. External supplier.

6. If the person or organization that receives your process output pays to purchase that output, we call that an:
   a. Internal customer.
   b. Internal supplier.
   c. External customer.
   d. External supplier.

7. Which of these would you NOT do in preparing for an upcoming customer visit?
   a. Clean up the area.
   b. Find out what the customer is interested in so you can be prepared to answer any questions.
   c. Coach the staff on effective customer relation interactions.
   d. Schedule high-impact meetings during the customer visit.

8. When it comes to quality, how good is good enough?
   a. 90 percent
   b. 95 percent
   c. 99 percent
   d. 100 percent is always your target

9. (*True or False*) As long as your product quality is good, that is all a process technician needs to worry about. Service quality is other people's job.

10. (*True or False*) If you focus on quality (product quality and service quality), you expect that profits will naturally follow.

11. (*True or False*) Customer complaints provide your business with a great opportunity to demonstrate your commitment to quality to your customers.

12. (*True or False*) Internal customers are just as important as external customers and should be treated the same.

13. (*True or False*) The golden rule is an excellent measure of how to behave when dealing with customers.

14. (*True or False*) Process technicians can reasonably expect to have direct contact with customers at some point in their career.

15. (*True or False*) Service quality and customer service interactions are not as important as product quality.

NOTE: Answers to Checking Your Knowledge questions are in the Appendix.

Student Activities

1. During the course of the next week, pay attention to customer service interactions at the grocery store, the department store, the hardware store, and so forth.
   a. Write a brief summary of an example of the worst customer service interaction that you experienced. What made this experience unpleasant? What can you learn from this experience and apply to your job as a process technician?
   b. Write a brief summary of the best customer service interaction that you have experienced. What made this experience pleasant? What can you learn from this experience to apply to your job as a process technician?
   c. One of the best ways to learn how to treat your customers with respect is to put yourself in their shoes. List at least five tangible ways that you can be a better customer when you go to the store.

2. Role-play positive and negative interactions that might occur with a customer coming to inspect a product or produce process. In small groups, discuss what went right or wrong and come up with practical ideas for unexpected interactions.

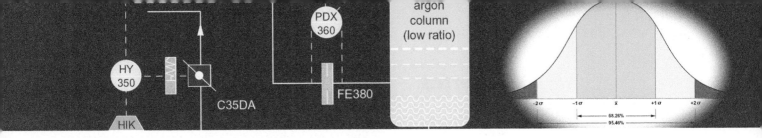

Chapter 4

Quality Management Systems—International Standards (ISO)

"Quality is never an accident. It is always the result of intelligent effort."

~ JOHN RUSKIN

 ## Objectives

Upon completion of this chapter you will be able to:

4.1 Explain the purpose of the ISO 9001 quality management system standard. (NAPTA Quality, Course Overview, 7, 8 and Variance and Operating Consistency 7) p. 38

4.2 Describe the key elements of ISO 9001 that apply to process technicians. (NAPTA Quality, Continuous Improvement and Corrective/Preventive Action 4, 6, 7) p. 40

4.3 Explain the purpose of the ISO 14001 standard and compare it to the ISO 9001 standard. (NAPTA Quality, Continuous Improvement and Corrective/Preventive Action 9*) p. 43

4.4 Define appropriate behaviors to exhibit in order to maximize the effectiveness of an audit. (NAPTA Quality, Course Overview 9) p. 45

4.5 Describe the purpose and criteria of the Malcolm Baldrige National Quality Award. (NAPTA Quality Course Overview 6) p. 49

*North American Process Technology Alliance (NAPTA) developed curriculum to ensure that Process Technology courses will produce knowledgeable graduates to become entry-level employees in process technology. Objectives from that curriculum are named here in abbreviated form. For example, "(NAPTA Quality, Continuous Improvement and Corrective/Preventive Action 9) means that this chapter's objective 3 relates to the NAPTA curriculum objective 9 on continuous improvement.

Key Terms

Accreditation—the formal recognition by an independent body, generally known as an accreditation body, that a prospective organization operates according to international standards, **p. 39**

Audit—a review of the management system to see whether the management system is effectively implemented, **p. 39**

Auditee—an entity or person being audited, **p. 46**

Auditor—a person conducting an audit, **p. 45**

Documentation—written or electronic information that records actions or defines how a task is to be accomplished, **p. 39**

International Organization for Standardization (ISO)—federation made up of over 160 member countries that have agreed to publish common standard methodologies for various aspects of conducting global business, **p. 38**

ISO 9001—the internationally recognized quality management system standard, **p. 39**

ISO 14001—the internationally recognized environmental management system standard, **p. 43**

ISO registered—the recognition granted to a company by an accredited registrar when the registrar has verified that the company has effectively implemented the management system, **p. 39**

Malcolm Baldrige National Quality Award (MBNQA)—a U.S.–based award established by Congress in 1987 that recognizes not only the effectiveness of a management system but also the company results achieved as a result of its management system, **p. 49**

Management system—a collection of activities, usually but not always documented, that are intended to ensure that products and services meet specified needs, **p. 38**

Nonconformance—evidence of deviation from documented requirements, **p. 40**

Records—evidence that a task has been accomplished, **p. 41**

4.1 Introduction

Management system a collection of activities, usually but not always documented, that are intended to ensure that products and services meet specified needs.

A **management system** is designed to ensure quality services and products. If you think back to Chapter 1, *Introduction to Process Quality*, you will remember that the definition of quality was hard to pin down because quality is in the eye of the beholder. The "needs" that a management system is intended to fulfill are also somewhat hard to pin down. There are the needs of the customer, the needs of management, the needs of various governmental regulatory agencies, and many others.

In a typical processing industry plant, it is not a matter of choosing which needs to meet, but of trying to meet all the needs. As a process technician, you will absolutely be impacted by and involved in the management system of your company in some way or another. You will certainly be expected to follow the company's management system. You may be asked to participate as an auditor, verifying that other parts of the management system are effectively implemented. You may also be audited by an external party whose job it is to verify that your part of the management system is implemented effectively.

As our economy has become more global, it has become increasingly important that the systems and protocols among countries be standardized. An organization called the **International Organization for Standardization (ISO)** has established such a set of standards.

International Organization for Standardization (ISO) federation made up of over 160 member countries that have agreed to publish common standard methodologies for various aspects of conducting global business.

The U.S. representative to the ISO is the American National Standards Institute (ANSI). ANSI, in conjunction with the American Society for Quality (ASQ), publishes the ISO standards in the United States. Some examples of the standards published include the ISO 9001 quality management standard and the ISO 14001 environmental management standard, both of which are commonly implemented in process industries.

Discussion in this chapter covers the pertinent portions of ISO 9001 and ISO 14001, including a comparison of these two management systems. It will concentrate specifically on the elements of these standards that apply to the day-to-day life of a process technician.

The role of a process technician as an auditor and as an auditee is also covered here. Finally, there is a brief discussion about the Malcolm Baldrige National Quality Award (MBNQA). Some companies are now moving toward the MBNQA criteria as a way of going beyond the "basic" requirements of ISO 9001.

What Is ISO 9001?

The International Organization for Standardization or **ISO 9001** is the internationally recognized quality management system standard (Figure 4.1). This management system has been implemented by thousands of companies around the world. Many customers require their suppliers to achieve and maintain ISO 9001 registration. ISO registration is a process whereby a company hires an outside company, called a registrar, to **audit** its management system against the standard. To become **ISO registered**, the registrar must verify that the company has effectively implemented the management system. If the registrar deems that a company has effectively implemented the entire management system, it will grant a certificate of registration.

ISO 9001 the internationally recognized quality management system standard.

Audit a review of the management system to see whether the management system is effectively implemented.

ISO registered the recognition granted to a company by an accredited registrar when the registrar has verified that the company has effectively implemented the management system.

Figure 4.1 Quality management systems cover a variety of quality concepts.

Registrars (ISO uses the term *certification bodies*) are *accredited* to grant registration that a company has implemented the standard. There is usually only one **accreditation** body for each country. In the United States, accreditation is managed by the Registrar Accreditation Board.

The ISO 9001 management system is made up of 10 elements (or clauses) with subelements (subclauses), each of which contains requirements that must be effectively implemented. In this sense, the management system is like an umbrella covering many individual topics.

The registrar's audits are done in two stages. Stage one evaluates an organization's readiness for a stage two audit. It can often be done remotely. Stage two is always done on site. The registrar will interview employees and review **documentation** to determine whether the organization is meeting the standards. The initial registration will cover every element of the standard. Subsequent surveillance audits are usually performed annually to maintain the registration. The registrar will have to cover all the elements of the audits every 3 years but will not necessarily cover every element during every annual audit.

The first international standard for a quality management system was issued in 1987, published in the United States as ANSI/ASQC Q92-1987. The first update to this initial standard was in 1994 and was called ANSI/ASQC Q9001-1994 and was considered a minor revision. The next revision, in 2000, was major. A second minor revision in 2008 sought to

Accreditation the formal recognition by an independent body, generally known as an accreditation body, that a prospective organization operates according to international standards.

Documentation written or electronic information that records actions or defines how a task is to be accomplished.

clarify issues raised in application of the 2000 edition. The version issued in 2015 is called ANSI/ISO/ASQ(E) Q9001 and is considered a second major revision. The ISO attempts to review each of its own standards every 5 years.

Changes made for the ANSI/ISO/ASQ(E) Q9001 are summarized as:

- Greater focus on the customer
- Policies and objectives aligned with the strategies of the organization
- Much more emphasis on risk-based thinking
- Greater flexibility regarding documentation.
- Increased requirements for involvement of higher management.

The elements and sub-elements of the current ISO 9001 quality management system standards can be found on the ISO website.

4.2 Key Elements of ISO 9001 Quality Standards

Many of the elements of ISO 9001 have a minimal impact on the role of a process technician; other elements are the responsibility of the process technician. The key elements that affect process technicians are listed first. In addition, we provide a brief description of some of the clauses of ISO 9001. If there comes a day when you are in management, you will be responsible for the effective operation of the entire quality system.

Key Elements of the ISO 9001 for Process Technicians

Competence, training, and awareness—Requires procedures for identifying training needs and providing training for all personnel to meet those needs.

Control of monitoring and measuring equipment—Requires procedures for selection, control, calibration, and maintenance for measuring test equipment.

Identification and traceability—Requires procedures for identifying product during all stages of production (Figure 4.2), delivery and installation, and individual product or batch-unique identification as needed.

Figure 4.2 Procedures are in place to enable a company to trace each product and each batch it produced.
CREDIT: Courtesy of Willie L. Myles.

Product Traceability

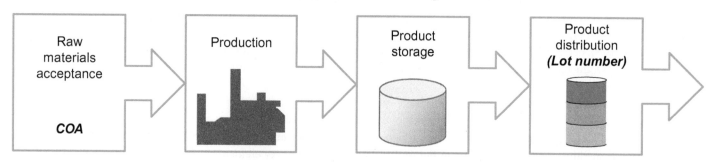

Monitoring and measurement of processes—Requires procedures to ensure that production and installation processes are carried out under controlled conditions, which include documentation, monitoring, and control of suitable process and product characteristics, use of approved equipment, and criteria for workshop.

Monitoring and measurement of product—Requires that markings, stamps or labels be affixed to product throughout production and installation to show conformance or **nonconformance** to tests and inspections.

Nonconformance   evidence of deviation from documented requirements.

Documentation requirements—Requires establishing and maintaining procedures for controlling documentation through approval, issue, change, and modification. Requires procedures for identification, collection, indexing, filling, and storage of quality **records**.

Records evidence that a task has been accomplished.

Overview of ISO 9001 Standards

Some elements of ISO 9001 are the responsibility of process technicians and some have a minimal impact on the technician's role. The following section is a brief description of selected clauses that will affect process technicians.

ISO 9001, 3. TERMS AND DEFINITIONS   Some terms and definitions do not vary between one industry and another. However, some may have very specific and unique meanings in context of a particular field. It is important for process technicians to become familiar with the terms and definitions specific to their field and their facility.

ISO 9001, 5. LEADERSHIP   All three sub-elements within clause 5 define leadership responsibilities, and management is required to be more involved in ISO quality review and control than previously. Management must focus on quality and customers, establish a quality policy, and define quality management system (QMS) roles and responsibilities. These activities may affect the process technician, but the direct responsibility lies with leadership, usually management.

ISO 9001, 6. PLANNING   The first of the three sub-elements in the Planning clause affects the process technician the most. It focuses on risk-based thinking. In planning, the process technician and others will look at identified risks in terms of what the risks are, who addresses them, how to address them, and when to address them. Planning is *proactive* to prevent or reduce the need for corrective actions later.

The second sub-element in Planning focuses on the objectives of the QMS, ensuring that they are measurable, monitored, communicated, aligned to the policy of the management system, and updated when needed.

The final sub-element states that any change to the QMS must be carefully considered before being adopted.

ISO 9001, 7. SUPPORT   After planning, the organization needs to determine what is needed to meet its objectives. The organization needs to:

- Determine and provide the resources needed for the establishment, implementation, maintenance, and continual improvement of the management system.
- Determine the necessary competence of persons doing work that affects the organization's QMS.
- Ensure that these persons are competent based on appropriate education, training, or experience.
- Take actions to ensure workers acquire the necessary competence and evaluate the effectiveness of the actions taken.
- Retain appropriate documented information as evidence of competence.

These sub-elements have their own sub-elements. Much of the work of the Support step will involve the process technician. Resources need to be provided by top management, but the process technician may be tasked with communicating what those resources are. Some of the sub-elements under resources are infrastructure, environment, people, and monitoring and measuring resources.

Sub-element 7.2 addresses competence of personnel. Process technicians will be affected by this sub-element, though it is management's responsibility to evaluate its employees' ability to meet the objectives of the QMS and to take action to address any areas in which personnel are lacking.

Sub-element 7.3 is especially pertinent for process technicians. It requires them to be aware of policies, objectives, and how these elements affect the QMS, as well as the benefits of improving quality performance and the consequences of not meeting requirements. Quality awareness must be focused on meeting both customer and regulatory requirements.

Sub-elements 7.4 and 7.5 focus on communication and documentation. It is up to the organization to determine what, when, who, with whom, and how communication relevant to QMS will be handled. This clause requires that the sequence and interaction of QMS processes be determined. This includes very basic communications, like input and output between parts of the process. It also includes deciding methods of communication for internal and external purposes. Documentation for ISO is less structured than in the past and more oriented to the shape and needs of the specific organization. Unlike past versions, the latest standard, ISO 9001:2015, does not mandate any required procedures or manual; AS9100D and ISO 9001:2015 mandate only a quality manual and one procedure for control of nonconforming outputs. The process technician is involved in the feedback on methods and effectiveness from every area of the process.

ISO 9001, 8. OPERATIONS   Sub-element 8.1 addresses developing processes. This clause is about operational planning and control. The organization needs to set up supplier accounts; purchase supplies, tools, and vehicles; train employees on the process; ensure adequate staffing; and clearly communicate the instructions on procedures.

Sub-element 8.2 is about determining requirements. This clause includes communicating to customers what you are selling, how they can expect to be dealt with, how you will treat their property, and what will happen if there are problems. Obtaining customer views and perspectives is part of this subsection. Also, the organization needs to determine the requirements it will need in order to sell products and services. This will consider legal and industry norms, as well as what the organization determines its own needs are. In addition, the organization will review its ability to provide what it is selling. This will consider customer orders, requirements of the facilities, supplies, and legal and industry standards. This clause also addresses what happens when a customer's requirements or the procedures change. Documentation of change is addressed here.

Sub-element 8.3 is to carry out the design. The intent of this subclause is that organizations will plan and control design and development projects to ensure that the project's requirements can be met within the time and budget allotted. Determining the stages of the project; the inputs required to meet standards, codes, and other requirements; responsibilities, authority, and interfaces are parts of this sub-element. Delineating the desired results, planning design reviews, and verifying that design inputs meet input requirements are other activities covered here. Documentation is a process that will need to be designed to fit the needs and structure of the particular organization.

Sub-element 8.4 is to monitor suppliers. Organizations must be sure that resources that are supplied externally will meet the requirements. Determining and applying criteria for these providers is part of this sub-element. This requires providing the external provider with the organization's requirements. Ultimately, the organization is responsible for the goods and services it passes along to its customers.

Sub-element 8.5 is about managing production. This subclause lists control requirements for the organization to use, as applicable. Process technicians are involved in identifying processes and interactions between processes, using the QMS to control operations, to schedule operations to meet delivery requirements, and to define and document operating scheduling processes and their interactions with logistics processes. Creation of flow charts, instructions, and other documentation for employees, and performance indicators is described in this subclause. Also covered here are testing, validation of customer approval, change management, and production-related indicators.

Sub-element 8.6 is about controlling the end product. This subclause states that the organization will have a process in place to monitor and measure the characteristics of the product to ensure requirements are being met. The auditor will check that records are maintained that show conformity. Products must meet the requirements laid out in 8.1 before being released. Process technicians are key people in seeing that this step is followed.

Sub-element 8.7 addresses documentation of actions related to nonconforming products. The organization needs a documented procedure describing how nonconforming products are detected, handled, and controlled (Figure 4.3). These products may be categorized as nonconforming because they do not meet customer requirements or regulatory requirements. The reaction of the organization to these nonconforming products includes segregating them, notifying the customer and/or government agency, and short-term corrective action. Documentation is important; everyone who will be affected needs to know what methods are to be used in these situations.

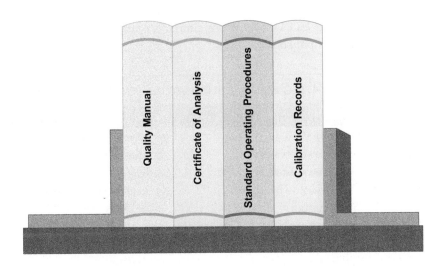

Figure 4.3 Organizations need documented procedures.

Sub-element 8.9 focuses on monitoring, measuring, analyzing, and evaluating the QMS. This requires a program for internal audits at planned intervals. It also includes customer satisfaction and testing results to judge the effectiveness of the organization's QMS.

Sub-element 8.7 addresses improvement. Improvement is based on identifying opportunities to change the organization's processes, products, and services to enhance customer satisfaction. Other sub-elements refer to controlling nonconformities, taking corrective actions, and enhancing the effectiveness of the QMS.

4.3 What Is ISO 14001?

ISO 14001 is an environmental management system standard. It differs from ISO 9001 in that **ISO 14001** is specifically intended to provide process guidance for environmental management systems. ISO 14001 is organized as follows:

ISO 14001   the internationally recognized environmental management system standard.

1 Scope
2 Normative references
3 Terms and definitions
. . .
4 Context of the organization
. . .
4.4 Environmental management system
5 Leadership
. . .
5.2 Environmental policy
6 Planning
. . .
6.2 Environmental objectives and planning to achieve them
7 Support
7.1 Resources
. . .

8 Operation
9 Performance evaluation
10 Improvement

Comparison of ISO 9001 and ISO 14001

As stated, ISO 9001 is a quality management system (QMS) which gives organizations a systematic approach for meeting customer objectives and providing consistent quality. ISO 14001 is an environmental management system (EMS) which gives organizations a systematic approach for measuring and improving their environmental impact.

ISO 9001 and ISO 14001 share a similar structure and have many similarities. See Table 4.1 for a comparison.

Table 4.1 Comparison of ISO 9001 and ISO 14001 Requirements

ISO 9001	ISO 14001
1 Scope	0.5 Contents of this international standard
2 Normative references	2 Normative references
3 Terms and definitions	3 Terms and definitions
4 Context of the organization	4 Context of the organization
5 Leadership	5 Leadership
6 Planning	6 Planning
7 Support	7 Support
8 Operation	8 Operation
9 Performance evaluation	9 Performance evaluation
10 Improvement	10 Improvement

The main level of standards for 14001 is organized in the same way as those for 9001, but sub-elements have specific reference to environmental concerns. They also are organized along the lines of the plan–do–check–act (PDCA) cycle.

Because of the similarities between these two management system standards, many companies combine them into one system when they are creating their quality management plans. If you were to look at other elements of the standards, you would find even more overlap. Both standards require corrective and preventive action processes, both require internal auditing, both require management responsibility, and so on. Many companies now implement an integrated management system that follows the ISO 10011: 2018 Guidelines for Quality and/or Environmental Management Systems Auditing standard.

Your company may employ these integrated guidelines. It may have its own harmonized management system that is designed to meet the requirements of both ISO 9001 and ISO 14001. Depending on the company you work for and the types and uses of the chemicals it produces, you may see many other management systems as well. Examples of these other management systems are listed in Table 4.2.

Table 4.2 Other Quality Management Systems Technicians May Encounter

Management System	Application
ISO 17025 Testing and Calibration Laboratory standard	Laboratories
Food and Drug Administration Good Manufacturing Practices	Drug chemicals (active ingredients)
Good Manufacturing Practices Guide for Bulk Pharmaceutical Excipients	Chemical excipients (inactive ingredients)
Bureau of Alcohol, Tobacco, and Firearms standards	Ethanol products
Supervised by competent rabbi and certified as kosher with a hechsher (a symbol or statement of compliance)	Kosher applications

As you can see, depending on which company in the processing industry you work for, and which business unit within the company, you may be required to follow a management system that is traceable to a number of different standards. In the case of quality, failing to follow the guidelines is bad because it leads to poor-quality product and potentially dissatisfied customers. In the case of drug products, following the management system is required to protect the lives of the people who will consume your chemicals.

4.4 The Process Technician and Audits

One requirement of nearly all management systems is that of internal audits (Figure 4.4). The old ISO 10011 quality audit standard, which still applies today, defined an audit as "a systematic examination of the acts and decisions by people with respect to quality to independently verify or evaluate and report degree of compliance to the operational requirements of the quality program, or the specifications or contract requirements of the product or service." In other words, auditing is examining evidence to demonstrate conformance to requirements.

Figure 4.4 A. and **B.** Internal audits are part of most management systems.
CREDIT: A. ID129614729:Shutterstock. **B.** michaeljung/Shutterstock.

A.

B.

Notice that the purpose is not to find fault but to demonstrate conformance. Process technicians can make very good auditors.

Let us describe what an **auditor** does and then you can see how a process technician could fill this role. A management systems audit can be broken down into two main questions:

1. Does the management system meet the requirements of the standard?
2. Is the management system implemented effectively?

Auditor   a person conducting an audit.

The design and setup of the management system is probably already in place, so the first part of the audit is relatively easy. The second part is where the rubber meets the road; essentially, are you doing what you said you would do? The auditor will look at your procedures, then look at what you actually do, and then will determine whether you are doing what you are supposed to be doing.

Auditors will interview the people working the floor and ask questions about what they do. They will look at the records of what happened yesterday, last week, and last month to see whether it can be demonstrated that the process is being managed according to the requirements of the management system.

The kind of things auditors look at depends on the element of the standard that is being audited (Figure 4.5). Auditors will look at process logs, transfer sheets, inventory records, training records, calibration records, and lab analyses. They will check to see whether the document control procedures are being followed. Are there any unauthorized copies of the procedures lying around? Are there any unauthorized changes made?

Figure 4.5 Auditors at work.

CREDIT: AB Forces News Collection/Alamy Stock Photo.

Auditors will validate that the record retention rules are enforced according to the management system requirements. An auditor must be independent of the area being audited, so you cannot audit within your own department, but you can audit another manufacturing unit within your plant. By using process technicians from one area to audit other areas, companies can take advantage of the process technicians' detailed knowledge of how things are *really* done to drive continuous improvement throughout the organization.

The Process Technician As an Auditee

Even if you do not serve as an auditor, you will almost certainly be audited at some time in your tenure as a process technician. Unfortunately, some people struggle with being audited. It makes them uncomfortable and, in some cases, seriously anxious. However, there are ways to prepare. One major way to prepare is to be trained as an auditor. Such training makes you a much better **auditee**, because you know exactly what the auditor is looking for and you can provide answers in a timely and efficient fashion. Understanding what auditors look for will also help you in the job you perform. Table 4.3 provides some common scenarios that you will encounter as an auditee and tips on how to respond positively.

Here is a wrap-up of this topic with an outline of audit considerations for an auditee:

Auditee   an entity or person being audited.

1. Ethical Considerations—ethical behavior is almost certainly a corporate requirement. Never put yourself, the plant, or the company in a compromising situation.

   a. Do

   - Answer every question honestly.
   - Answer only the questions that are being asked.
   - Narrow down the request to specifics.

   b. Don't

   - Lie to an auditor.
   - Volunteer information or opinions.

Table 4.3 Responses to Auditing Scenarios

Audit Scenario	How to Respond Positively
Auditors ask questions and notes.	Answer the question that is asked as completely and succinctly as possible. Never volunteer information that was not requested. Be polite, but do not become a nervous chatterbox.
Auditors may discuss nonbusiness-related topics as a way of making you feel more comfortable.	Enjoy it. If they have an hour to spend with you and they take 30 minutes to discuss horses or cars or boats, that just leaves less time for other questions. However, when an auditor does ask a business question, do not stall. It is okay for the auditor to divert the auditee's attention, but it is not okay for the auditee to stall and attempt to divert the auditor's attention.
The auditor asks you about various parts of the management system.	You need to know your management system procedures inside out. This does not mean knowing everything about the company, but everything that has a direct impact on your specific job.
The auditor asks you for information that is outside of your area of responsibility.	Do not let this fluster you. The auditor may not have a clear understanding of your area and may just be fishing for information. It is okay to say that you do not have that information and that the auditor will have to seek that information elsewhere. If you know where to send the auditor, then help by directing the auditor to the right person.
The auditor asks a question you do not understand.	Ask the auditor to clarify. Do not attempt to respond if you do not know the answer. Make sure you understand the question before you attempt to answer.
The auditor asks you how to perform a certain task.	Tell the auditor that you follow the documented procedure and then show the auditor the written procedure.
The auditor presses you to say how you do the task without looking at the procedure.	Tell the auditor that, in order to ensure you are following the most current version of the procedure, you would never just perform the task without first reviewing the procedure. Do not fall for this trick.
Upon your answering a question, the auditor makes a few notes, then looks up at you with a questioning look and waits expectantly.	Return the auditor's gaze and sit patiently, awaiting the next question. Auditors are trained to use silence as a tool to get people talking.
The auditor asks to see a certain task being performed.	If the task is going to be performed anyway, then allow the auditor to watch. However, there is no requirement that you (or anybody in the company) perform unnecessary tasks just to satisfy an auditor's question.
The auditor asks to perform some part of the task.	Politely decline. An auditor is to look and listen but not touch. In fact, one of the roles during an audit is to keep the auditor safe. Remember that this is your unit, not the auditor's, and he or she may not recognize all the hazards that are familiar to you.
The auditor asks you a question, and you do not know the answer.	Answer truthfully that you do not know the answer. Never try to pretend or bluff. The auditor will see through it every time.
The auditor discovers a nonconformance in your area.	Acknowledge the audit finding and thank the auditor for pointing out an opportunity for improvement. Make a note of the issue so you can follow up with the right person after the audit. This response demonstrates an attitude of continuous improvement and will be received positively. Unless directed to do so by your supervisor or unless you see an obvious safety hazard that requires immediate response, do not attempt to address the audit finding on the spot with the auditor watching. This might give the auditor the impression that you are fixing the problem without giving the matter serious thought. Just accept the finding and move on.
The auditor asks to go off into the unit alone to look at something.	Politely decline. For safety reasons, auditors should never be unescorted.

- Point the auditor to someone else's deficiencies.
- Air dirty laundry.
- Attempt to "bluff" the auditor.
- Alter data.

2. Social Considerations—contrary to popular belief, auditors are people too. Their perception is 90 percent of their reality. Consequently, how they perceive your performance will influence how they document their audit findings.

   a. Do
      - Be friendly.
      - Be careful about joking—it could backfire.
      - Be on your best behavior.
      - Treat the auditor with respect.
      - Take notes—show you are interested.
      - Watch your body language.

    b. Don't

- Joke around unnecessarily.
- Leave auditor alone.
- Fight with auditor.
- Be defensive or argue—if there is a disagreement, the auditee should refer the auditor to another company official.
- Whine or complain.
- Panic—it is okay to ask for help if you are not sure of the auditor's question.

3. Technical Considerations—the auditor's perception of your area will initially be based on you personally, then on your technical knowledge, so technical accuracy is a must.

    a. Do

- Project the following to an auditor during the review:
  1. Good system in place
  2. Well organized
  3. Knowledgeable personnel
  4. Safeguards against single point failures
- Look it up.
- Ask for clarification.
- Know your area well.
- Ask to see the requirement in question.
- Use the resources around you.
- Your homework—be familiar with procedures, records, and manuals.
- Admit it when you do not understand or do not know the answer.

    b. Don't

- Fumble around looking for things.
- Offer documents that have not been requested.
- Give the auditor entire files.
- Come to close-out meeting to debate issues.
- Fix issues immediately unless instructed to do so by a supervisor.

4. Personnel Considerations—sometimes the best answer to a question is, "Let me get the right person to answer that question for you." You should know who is likely to have information about different areas.

    a. Do

- Steer the auditor to the right people.
- Treat the auditor with respect.
- Recognize that the auditor is here to help.
- Listen to the auditor. It is a sign of respect and interest.

    b. Don't

- Try to answer every question yourself.
- Point blame at somebody else.
- Demean the auditor's position.
- Ask the auditor to explain an internal or regulatory requirement.

5. Location Considerations—where the audit is conducted is crucial to the effectiveness of the audit. A good auditor will want to go where the action is. A good auditee will want to stay in a nice safe conference room.

   a. Do
      - Meet in your office or a conference room.
      - Go get requested information and bring it back to the auditor.
      - Provide the entire file only if the auditor insists.
      - Know your area so you can take the auditor where he or she wants to go without wandering around.

   b. Don't
      - Meet with the auditor in your file room.
      - Wander around the unit just to see what you might see.
      - Send an auditor anywhere—escort the auditor there.

4.5 Malcolm Baldrige National Quality Award

The **Malcolm Baldrige National Quality Award (MBNQA)** was established by the U.S. Congress and signed into law by Ronald Reagan in 1987. Named for Malcolm Baldrige (Figure 4.6), who served as Secretary of Commerce from 1981 to 1987, this award was created to stimulate American companies to improve quality and productivity by recognizing not only the effectiveness of the management system but also the company results achieved as a result of the management system. It is the nation's highest presidential honor for performance excellence.

Malcolm Baldrige National Quality Award (MBNQA) a U.S.–based award established by Congress in 1987 that recognizes not only the effectiveness of a management system but also the company results achieved as a result of its management system.

Figure 4.6 Malcolm Baldrige.

CREDIT: Dennis Brack bb71/Alamy Stock Photos.

The ISO 9001 quality management system standard has almost become the baseline for most companies. To distinguish themselves in the marketplace, companies must continually push forward and try to outdo their competition. Implementing the criteria for the MBNQA is recognized as a lofty goal for the next level of quality and excellence.

Three MBNQA awards can be given annually in six categories:

1. Manufacturing
2. Service Company

3. Small Business

4. Education

5. Healthcare

6. Nonprofit

The management system for the companies that some of you will go to work for may include requirements based on the MBNQA criteria as well as many others that we have already discussed. The biggest difference between the MBNQA criteria and those of the other management systems that have been reviewed is the focus on results. ISO 9001 and ISO 14001 both focus on having the right work processes in place to ensure effective implementation. The MBNQA criteria require a demonstration of excellent results. The criteria are organized into seven categories as follows:

1. Leadership

2. Strategic Planning

3. Customer and Market Focus

4. Measurement, Analysis, and Knowledge Management

5. Workforce Focus

6. Process Management

7. Results

Did You Know?

Malcolm Baldrige was named Professional Rodeo Man of the Year in 1980 and was installed in the National Cowboy Hall of Fame in 1984. He died at age 64 as the result of a rodeo accident in 1987. He received the Presidential Medal of Freedom from President Ronald Reagan posthumously in 1988.

CREDIT: Moriz/Shutterstock.

Summary

You will encounter management systems of various types in your role as a process technician. Some may be quality management systems such as ISO 9001. Others may be environmental management systems, such as ISO 14001. An increasing percentage of management systems are integrated, meaning the company system covers more than one standard. These management systems document the processes that a company should have in place to drive continuous improvement into its operations. Both ISO 9001 and ISO 14001 are organized around the plan-do-check-act continuous improvement cycle. Process technicians have direct responsibility for maintaining portions of a quality system, which may include the following:

- Documentation requirements
- Competence, awareness and training
- Identification and traceability
- Control of monitoring and measuring equipment
- Monitoring and measurement of processes
- Monitoring and measurement of product

Keeping the procedures up-to-date and following those procedures to the letter are vitally important to produce a quality product safely and responsibly. If the procedures are wrong, you should work within the system to get them fixed, but you should NEVER arbitrarily think you have a better way and try it out to see what happens.

Additionally, nearly all process technicians will be involved in the audit function at some point as an auditee and perhaps as an auditor. As an auditee is important to understand the role you play in presenting a positive impression to the audit team. Remaining calm, having answers at your fingertips, and being a positive person help the audit team find the evidence of conformance that it is looking for. Due to

their grassroots familiarity with the process operations, process technicians make excellent additions to the audit teams. Conversely, audit training makes better auditees out of process technicians and day staff alike.

For those companies looking to go above and beyond the minimum requirements for a recognized quality management system, the Malcolm Baldrige National Quality Award provides criteria that not only drive excellence but also require measurement of the results in order to demonstrate excellence.

Checking Your Knowledge

1. Define the following key terms:
   a. Audit
   b. Auditee
   c. Auditor
   d. Documents
   e. ISO 9001
   f. ISO 14001
   g. Malcolm Baldrige National Quality Award
   h. Management system
   i. Nonconformance
   j. Records

2. (*True or False*) Both ISO 9001 and ISO 14001 require corrective and preventive action processes.

3. Which of the following are elements of the ISO 9001 quality management system standard that apply directly to the role of a process technician? (Select all that apply.)
   a. Competence, training, and awareness
   b. Emphasis on risk-based thinking
   c. Operational planning and control
   d. Identification and traceability
   e. Monitoring and measurement of product
   f. Monitoring and measurement of processes

4. ISO 14001 is intended to provide guidance in which area of business practice?
   a. Human resources
   b. Environmental management
   c. Housekeeping
   d. Maintenance

5. "Examining evidence to demonstrate conformance to requirements" is the definition of:
   a. A management system.
   b. An audit.
   c. A government inspection.
   d. An investigation.

6. Which **two** of the following answers are components of a management system's audit?
   a. Ensuring the management system meets legal requirements
   b. Ensuring the management system meets the requirements of the standard
   c. Ensuring the management system is effectively implemented
   d. Ensuring the management system is certified to national standards

7. Which of the following would not be reviewed during a quality audit?
   a. Medical records
   b. Transfer records
   c. Process records
   d. Calibration records

8. When it is okay to provide false information to an auditor?
   a. When the truth will get you in trouble
   b. When your supervisor instructs you to do so
   c. When you are not sure of the correct answer
   d. It is never okay to provide false information.

9. All the following are things you should NOT do with an auditor except:
   a. Joke about the workplace.
   b. Treat the auditor with respect.
   c. Argue with the auditor.
   d. Leave the auditor alone.

10. Which of the following is undesirable during an audit?
    a. Asking for clarification
    b. Redirecting a question to the appropriate party for an answer
    c. Using the resources available to you during the audit
    d. Looking around for things

11. All of the following actions are acceptable during an audit except:

    a. Pointing the blame at someone else

    b. Steering auditors to the right people

    c. Treating auditors with respect

    d. Listening to the auditor

12. (*True or False*) If possible, it is better to meet the auditors in a conference room rather than out in the workplace.

13. The national award for quality excellence in the United States is called the:

    a. Malcolm Baldrige National Quality Award

    b. National Quality and Productivity Council Award

    c. Presidential Freedom Award

    d. Deming Prize

14. (*True or False*) The ISO 14001 environmental management system standards are required by the EPA in all chemical plants in the United States.

15. (*True or False*) The ISO 9001 quality management system standard is voluntary for all process industry plants.

16. (*True or False*) Malcolm Baldrige was a United States Secretary of Commerce.

NOTE: Answers to Checking Your Knowledge questions are in the Appendix.

Student Activities

1. Have students go online to the ISO website and review the ISO 9001 standards and the ISO 14001 standards. Have them complete a two-column cross-reference of the main elements of 9001 and 14001 as structured below, plus the sub-elements they think might relate to them most directly in their role as process technicians. Discuss their choices of sub-elements.

Key Element of ISO 9001	Matching Element in ISO 14001
[9001 main elements here]	[14001 main elements here]

2. In small groups, have students (1) explain the following ISO elements, (2) discuss the purpose of each element, and (3) brainstorm ways a process tech can comply with each one. Use the following elements:

   a. Control QMS information

   b. Control of documents

   c. Identification and traceability

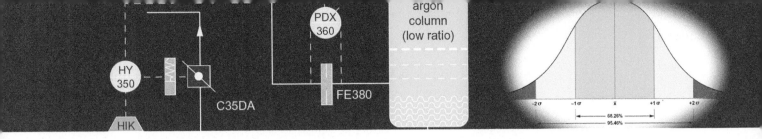

Chapter 5
Quality Management–Quality Reliability Planning

"Quality is never an accident; it is always the result of high intention, sincere effort, intelligent direction and skillful execution; it represents the wise choice of many alternatives."

~ WILLIAM A. FOSTER

 Objectives

Upon completion of this chapter you will be able to:

5.1 Define the three components of a quality reliability plan (QRP) and explain the value of using a QRP for process improvement. (NAPTA Quality, Total Quality Management 1*) p. 54

5.2 Explain the concept, the components, and the way to construct a quality design plan (QDP). (NAPTA Quality, Variance and Operating Consistency 14–16; NAPTA Continuous Improvement and Corrective/Preventive Action 3, 4, 6) p. 56

5.3 Explain the concept, the components, and the way to construct a quality control plan (QCP). (NAPTA Quality, Variance and Operating Consistency 14–16; NAPTA Continuous Improvement and Corrective/Preventive Action 3, 4, 6) p. 57

5.4 Explain the concept, the components, and the way to construct a failure mode and effects analysis (FMEA). (NAPTA Quality, Variance and Operating Consistency 14–16; NAPTA Continuous Improvement and Corrective/Preventive Action 3, 4, 7) p. 59

*North American Process Technology Alliance (NAPTA) developed curriculum to ensure that Process Technology courses will produce knowledgeable graduates to become entry-level employees in process technology. Objectives from that curriculum are named here in abbreviated form. For example, "(NAPTA Quality, Total Quality Management 1)" means that this chapter's objective 1 relates to the NAPTA curriculum objective 1 on total quality management.

5.5 Describe steps in the QRP process. (NAPTA Continuous Improvement and Corrective/Preventive Action 3, 4, 6, 7) p. 63

Key Terms

Failure mode and effects analysis (FMEA)—the third and final step in the quality reliability planning process, in which you examine every possible failure mode for the process in order to understand the consequences of process failures from the customers' perspective, **p. 59**

Quality control plan (QCP)—the second step in the quality reliability planning process. Process control parameters are specified to ensure the process is capable of meeting the needs of the customer. Process capability is captured at this stage, **p. 57**

Quality design plan (QDP)—the first step in the quality reliability planning process. The needs of the customer are translated into specific requirements (specifications), and the measurement techniques are specified to ensure the needs are met, **p. 56**

Quality reliability planning (QRP)—a procedure designed to ensure that when a new product is being introduced, proper communications occur between all of the functional groups and that a system is in place that allows the product to meet the customers' expectations consistently. The procedure consists of the quality design plan, the quality control plan, and the failure mode and effects analysis, **p. 54**

Risk priority number (RPN)—a calculated number within the failure mode and effects analysis process that helps prioritize the actions that need to be taken to minimize the risks of process failures and consequences to the customer, **p. 60**

Quality reliability planning (QRP) a procedure designed to ensure that when a new product is being introduced, proper communications occur between all of the functional groups and that a system is in place that allows the product to meet the customers' expectations consistently. The procedure consists of the quality design plan, the quality control plan, and the failure mode and effects analysis.

5.1 Introduction

Quality reliability planning (QRP) is the process that begins when a new product is introduced. It is designed to ensure that the customer will be satisfied with the product and service it receives (Figure 5.1). The quality design plan (QDP), quality control plan (QCP), and failure mode and effects analysis (FMEA) are pieces of this larger process that process

Figure 5.1 Failure mode and effects analyses are built on a quality control plan and quality design plan.

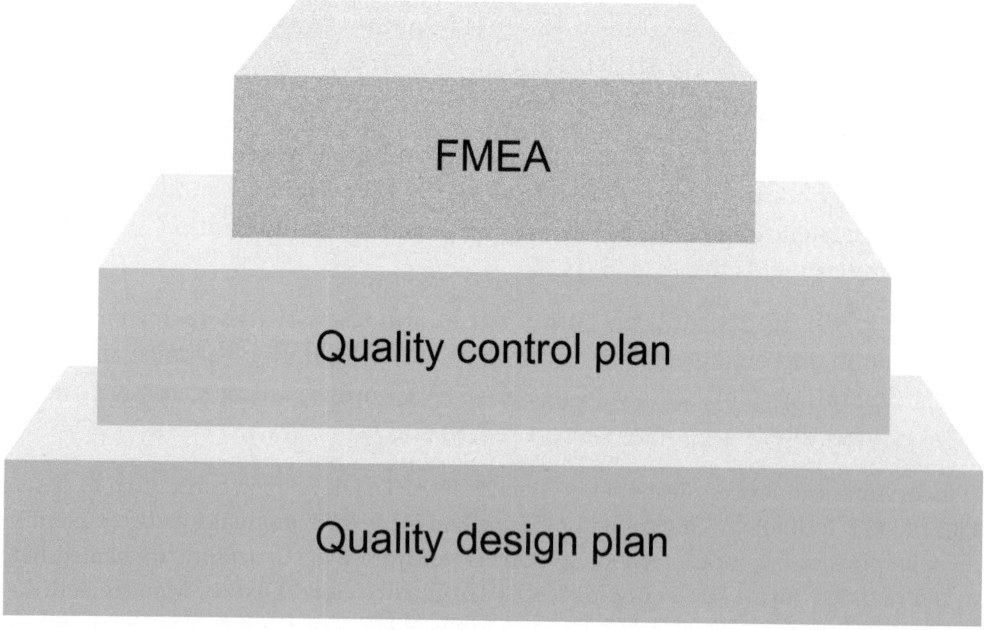

technicians will take part in. The benefit of using these tools has been documented by a variety of companies.

In the grand scheme of this textbook, quality reliability planning is another improvement strategy to employ. For your purposes, this chapter will review the entire procedure, but the discussion concentrates on the sections most applicable to process technicians.

Over the years, many businesses have adopted the use of the QRP tools as a process improvement technique. QRP has been designed to be specifically beneficial to the processing industries. It incorporates changes made to general industry procedures as a result of experiences in applying these tools.

QRP helps to overcome many problems that occur due to lack of communication between the product designers and the product producers. Industry has found that QRP better enables companies to manufacture products that can consistently meet and exceed the customers' expectations. QRP has resulted in better initial system designs, which have significantly reduced problems after a product goes into commercial production. QRP consists of three main components, each of which serves a useful and complementary purpose in the overall scheme of process improvement.

The basis of the QRP is the quality design plan (QDP). This plan is used to document customer expectations, translate those expectations into product specifications and methods, and analyze those specs and methods to ensure that they are capable of complying with the customer's performance requirements. The quality control plan (QCP) builds upon this foundation. It documents the manufacturing procedures for handling nonconforming product, describes the process parameters that control the quality of the product, defines the acceptable variation in the process parameters, and outlines the action steps to correct a process that is exhibiting unacceptable variation.

Finally, the failure mode and effects analysis (FMEA) is used to predict potential failures of the product, document the effects of these failures on the customer, and brainstorm potential causes for each failure. The FMEA is then used to generate a risk priority number (RPN). In cases in which an unfavorable RPN is calculated, the opportunity is provided to recommend process improvements.

QRP is not intended to be an individual activity. It is best applied as a team effort (Figure 5.2), although much of the initial work can be done in small groups to make better use of everyone's time. In addition to step-by-step instructions for applying this tool, charts will be provided outlining the sequence of events necessary to complete a QRP. On the charts, suggested primary leadership responsibilities are identified. However, other interested parties may be involved at any or all steps in the process. A suggested team membership roster would include representatives from the following:

Plant laboratory	LAB
Analytical measurements R&D	ANL
Applications R&D	APP
Process R&D	PROC
Production Department	PROD

A representative from the quality function, such as a quality engineer or quality manager, would also be included to serve as a facilitator of the QRP process and possibly act as a team scribe, recording ideas and decisions.

The rest of this chapter provides a column-by-column, form-by-form instruction guide to using QRP. Keep in mind that the underlying purpose of this tool is process improvement. The technical details of the forms are only important as an aid to process improvement. Blank forms for the quality design plan, the quality control plan, and the FMEA are included in instructor resources for the chapter. On the forms, you will see that each column is numbered. These column numbers will be used in the body of this guide as a reference for the instructions.

Figure 5.2 Quality reliability planning works best when applied as a team effort.

5.2 Quality Design Plan

Quality design plan (QDP) the first step in the quality reliability planning process. The needs of the customer are translated into specific requirements (specifications), and the measurement techniques are specified to ensure the needs are met.

Quality reliability planning begins with the **quality design plan (QDP)**. Customers describe their needs, which research and development (R&D) write as requirements. This step also identifies ways to measure that customers' specifications will be met. Its primary contribution to the overall process is in building an information base, which greatly increases the efficiency of the rest of the process. Because the type of information captured in the design plan is usually provided by R&D, the role of the process technician in this phase of the QRP is as an observer and/or as a process equipment specialist. Additionally, including new operators and technicians in this step is an excellent training opportunity for them to learn not only what needs to be done but also why it needs to be done. A thorough understanding of what the customer expects and how to measure the quality of the product increases everyone's awareness of the impact of the company's actions in making the product.

An example of the header section of the QDP is shown in Figure 5.3. The text under Column Instructions explains what is expected in each numbered column.

Figure 5.3 Example of the header section of a quality design plan.

Quality design plan for:_____   Page ____of _____   Date: _____

Customer	Acceptable limits	The quality attributes that control performance	Proposed limits for quality attribute	Detection method for the quality attribute (lab method)	Method capability, cm	If method capability is less than 2.50:			
						Recommended actions to improve precision	Respon-sible party	Actions taken	Date completed
Performance attribute 1	(If any apply) 2	3	4	5	6	7	8	9	10

Column Instructions

1–2. The first two columns are for documenting what exactly the customer needs the product to do in terms of its performance characteristics. In some cases, these may be the specifications for the product that your product is used to make. If possible, the customer should be asked to provide this information directly. Otherwise, the representative from Applications R&D will need to take the lead in defining the customers' expectations. Examples of end-product performance attributes might include the following:

- Vent collapse
- Cure rate
- Tensile strength
- Compression set
- Tear strength
- Lubrication

- Force to crush
- Rise time
- Solubility
- Volatility
- Surface tension
- Molecular weight distribution

3–4. In columns 3 and 4, Process R&D has the leadership role, working with Applications R&D. The purpose here is to translate the customers' expectations into meaningful product specifications. Limits for product specifications should be set to ensure complete compliance with the customers' needs. Examples of typical quality attributes include the following:

- Hydroxyl number
- Total polymer
- Acid number
- Viscosity

- Percent weight of water
- Alkalinity
- Purity
- Color stability

5–6. In columns 5 and 6, the detection method for the quality attribute is identified. Process R&D, in conjunction with the plant laboratory, should be able to describe the method and produce historical data on this method or similar methods illustrating the expected precision of the method. A laboratory method should exhibit variation no greater than 40 percent of the specification range (this equates to a Cm of 2.5, which you will learn about in Chapter 16 *Process Capability*) to ensure that your system is capable of meeting the specs. When this standard is not met, several options are available:

- Improve the precision of the method.
- Widen the specification band.
- Devise a sample plan, as an integral part of the production system, to compensate for the excessive variation.

7–10. For those analytical methods that can be improved, columns 7–10 are available to document the recommendations and to record the improvements. The analytical measurements group and the plant laboratory personnel are the most likely candidates for initiating this type of activity.

Did You Know?

Technically, the word *data* is plural, meaning pieces of information. The word for a single piece of information is *datum*. That is why we use phrases such as "the data are distributed" instead of "the data is distributed."

5.3 Quality Control Plan

The **quality control plan (QCP)** is the next step in the QRP. It is designed to document the manufacturing reaction to nonconformance and the exact critical in-process parameters that control the quality attributes that affect the customers' expectations. This is a key piece in the puzzle of process understanding. The plant manufacturing department representative has the lead role in the control plan, working closely with Process R&D.

Quality control plan (QCP) the second step in the quality reliability planning process. Process control parameters are specified to ensure the process is capable of meeting the needs of the customer. Process capability is captured at this stage.

Process technicians can and should play a key role in this phase of the QRP, as they are the ones who control the process on a day-by-day basis. It is common during these exercises for personnel who are not process technicians to have a different understanding of how the process is controlled than the process technicians do. In this case, one of two things will happen. Either you will help to change the current control scheme because of your input, or you will be convinced to change your behaviors based on a better understanding of what is important to the customer.

An example of the header section of the quality control plan is shown in Figure 5.4. The column numbers are explained in the text that follows.

Figure 5.4 Example of the header section of a quality control plan.

Quality control plan for:_____   Page ____of _____   Date: _____

Quality attributes	Sample plan				Specification limits for quality attribute	Corrective actions for non-conformance (outside limits)	Controlling process parameters	Acceptable limits for process parameters	Process parameter control scheme	Corrective actions for non-conformance (outside limits)
	Sample point	Freq.	No.	No.						
			Samples	Tests						
1	2	3	4	5	6	7	8	9	10	11

Column Instructions

1. In this column, all of the quality attributes that were identified through use of the quality design plan are reiterated. This column also captures any additional specifications that are included as double checks or any specifications that are indirect indications of quality.

2–5. These columns represent the results of any sample planning activities and provide documentation of the current sample schemes. The sample plan is a description of how many samples to take, from where, and possibly how to take them.

2. In column 2, enter the location(s) of where the samples will be taken. The list below provides examples of sample locations as well as shorthand terms for entering the information in the form.

ML—make line	PL—pipeline
MT—make tank	ST—storage tank
TT—tank truck	TC—tank car
DR—drum	CYL—cylinder

3. For periodic sample locations, such as line samples, input the frequency of sampling in column 3. For example, this might be every 4 hours, daily, hourly, and so on. For those samples that are not periodic, such as tank samples, enter "each" or "lot" as the frequency to indicate that each container is sampled.

4. If replicate samples are taken, in accordance with an established sample plan, then enter the number of replicate samples taken in column 4. For most of your products, this number will be a 1.

5. If replicate analyses are run on each sample, then enter the number of replicates in column 5. Many of your analytical methods routinely require duplicate analysis, to reduce the analytical variation.

6. In column 6, enter the current specification limits for the quality attributes either as a nominal value with the acceptable deviation or as a specification range with minimum and maximum values. Be sure to include the unit of measurement.

7. In column 7, document the current reaction scheme to nonconformance to the specifications. This piece of information is of utmost concern to your customers, who invariably bring up this topic in audits. Examples would include the following:

   - Isolate batch and rework
   - Destroy product
   - Sell as distress

8–9. In columns 8 and 9, you document the parameters in your process that actually control the quality of your product and how much variability is allowed in each one. These would be the physical control parameters of your process, such as the following:

   Reactor temperature: 125–135 degrees Celsius

   Column pressure: 40 PSI max

   Feed rate: 90.5 pph

   Transfer rate: 1500 GPM, nominal

   Mixing time: 16 hours, minimum

   Rate of addition: 1 drum in 30 minutes

10. In column 10, you must describe the system(s) used to control the process parameters. Process control schemes would include things such as the following:

   - Electronic controllers
   - Redundant RTDs
   - Administrative controls
   - Pneumatic controllers
   - D/P cells
   - Standard operating procedures (SOPs)

11. The last column of the control plan is where you document the actions to be taken when the process control parameters deviate from their acceptable limits. In many of the types of process controls in the process industry, these actions are things such as repairing malfunctioning controllers.

5.4 Failure Mode and Effects Analysis

The **failure mode and effects analysis (FMEA)** is the final step in the quality reliability planning process. FMEA is the portion of the QRP process that documents the effects of nonconformance on the customer and allows the team to brainstorm potential problems in the processes. This exercise is definitely a team effort.

All members of the team should have access to the quality design plan and quality control plan information for review prior to meeting to work on the FMEA. All team members should also have access to this instruction guide before attempting to complete the FMEA, in order to familiarize themselves with the ranking criteria listed in the following pages. The FMEA is applied here to quality failure modes. Process technicians may also find themselves part of a Hazard Operability Study, or HAZOP, team that is analyzing potential failure modes for safety and environmental issues. Although the mechanism is the same, the descriptions of the occurrence, severity, and detection sections may vary.

An example of the header section of an FMEA is shown in Figure 5.5. The text following the image explains what is expected in each column.

Failure mode and effects analysis (FMEA) the third and final step in the quality reliability planning process, in which you examine every possible failure mode for the process in order to understand the consequences of process failures from the customers' perspective.

Column Instructions

1. Column 1 begins the part of the QRP exercise in which you analyze every potential failure of your product to meet specifications. Some failures will be two-sided failures, such as total polymer, which can fail high or low. Other failures will be one sided, such as water contamination, which will only fail high. (Contamination has only a high limit. It cannot be lower than zero. The lower the contamination value, the better.)

Figure 5.5 Example of the header section of a failure mode and effect analysis.

FMEA for:_____   Page ____of _____   Date: _____

Give a description of the potential failure	Effect on the customer due to this potential failure	List all possible causes for each potential failure	Existing conditions						Resulting conditions					
			O C C	S E V	D E T	R P N	Recommended actions for RPNs >125	Respon-sible party	Actions taken	Date completed	O C C	S E V	D E T	R P N
1	2	3	4	5	6	7	8	9	10	11	12	13	14	15

In addition to specification failures, you should include any other pertinent, potential failure of your process that could impact the customer. Some examples of potential failures include the following:

- High hydroxyl
- Low hydroxyl
- Product not well mixed
- Contaminated when loaded
- Low total polymer
- High water
- Off color

2. For each failure mode documented in column 1, give a complete description of the effect(s) of that failure on the customer in column 2. Applications R&D will play a large part in this portion of the exercise. For example, a low total polymer failure may cause low loads in the customers' foam; a high hydroxyl number might cause vent collapse and shear collapse. This part of the QRP is your evidence that you have in-depth knowledge of how your product affects the customers' process.

3. In column 3, brainstorm all (within reason) possible causes for each of the failure modes listed in column 1. The generation of this list will involve the entire team. Consider historical information when generating the list. Typical causes in the past have been things such as the following:

- Failure to follow operating procedures
- D/P cell failure
- Pump failure
- Dirty tank truck
- Filter bag ruptured during loading
- Raw materials are off-specification
- Improper raw material ratio

NOTE: Columns 4 to 7 are used to analyze the system currently in place for its capability to achieve the desired outcome. There are three numerical rankings to assign: occurrence (OCC), severity (SEV), and detection (DET); and a **risk priority number (RPN)** to calculate. Each cause must have a separate set of rankings, because different failures may have different rankings depending on what caused the failure. An example of OCC would be the probability of a pump failure versus an airplane's falling out of the sky and landing on the equipment. Obviously, there is a higher probability that a given pump will fail than there is that an airplane will crash into the equipment. You must apply this thought process to each potential cause identified. Each column will be discussed separately.

Risk priority number (RPN)   a calculated number within the failure mode and effects analysis process that helps prioritize the actions that need to be taken to minimize the risks of process failures and consequences to the customer.

4. Column 4 assigns a numerical rank, on a scale of 1 to 10, which describes the probability of occurrence for each of the potential causes that have been documented. Because your ability to detect a failure is evaluated in a different column (6), you must narrow your

Table 5.1 Criteria for Assigning the Occurrence (OCC) Ranking

Rank	Criteria	Probability	Cpk
1	Remote probability of occurrence. Capability shows at least $x \pm 4\sigma$ within specifications.	1/16,000	1.33
2	Low probability of occurrence. Process in statistical control. Capability shows at least $x \pm 3\sigma$ within specifications.	1/2000	1.17
3		1/370	1.00
4	Moderate probability of occurrence. Generally associated with processes similar to previous processes that have experienced occasional failures, but not in major proportions. Process in statistical control. Capability shows more than $x \pm 2.5\sigma$ within specifications.	1/144	0.90
5		1/61	0.80
6	High probability of occurrence. Generally associated with processes similar to previous processes that have often failed. Process in statistical control, but capability shows $x \pm 2.5\sigma$ or less within specifications.	1/28	0.70
7		1/8	0.50
8	High probability of occurrence. In evaluator's view, failure is almost certain to occur.	1/4	0.40
9		1/2	0.20
10	System not in state of statistical control.	1/1	N/A

thoughts strictly to the probability of occurrence. Table 5.1 gives a description of what each ranking means. A probability of 1/370 means that with the current system, out of every 370 batches produced, 1 will fail. The use of historical data, including SPC/SQC type data, will be helpful. The ranking assigned may be based on data but sometimes represents a consensus opinion of the participants. If the issue in question is one that you have been tracking via control charts, then data would exist to calculate the Cpk.

5. Into column 5 goes a numerical severity ranking, on a scale of 1 to 10, of the expected severity of each identified potential cause. Applications R&D or the commercial organization may be best suited to describe the severity of each cause. In some instances, a particular failure mode may have the same severity ranking for each cause. In other cases, it is possible that a particular failure will have more severe consequences for some types of causes than for others. Table 5.2 lists the severity criteria.

6. Into column 6, you enter a DET ranking that describes your ability to detect a failure, once it occurs, before the product reaches the customer. You may either assign a DET rank based on a perceived probability of shipping a defect (using customer complaint data is even better) or use the measurement system capability (Cm). You will find discussion of this topic in Chapter 16, *Process Capability*.

7. Column 7 is for the risk priority number (RPN), which is the result of multiplying the OCC ranking by the SEV ranking by the DET ranking. The RPN can therefore be any number between 1 and 1,000. By using the RPN, you can determine which potential causes should be addressed first. A potential cause that has a moderate probability of occurring (5), causes a moderate severity situation with your customer (5), and has only a moderate probability of being detected before shipment (5) will yield an RPN value of 125.

8. A maximum RPN limit should be agreed upon by the team. Any potential causes having an RPN higher than the agreed-upon limit should have action plans to lower the RPN listed in column 8. A limit of 125 may work for most areas. The team may also want to set a maximum for the OCC, SEV, and DET categories separately. However, even if a potential cause has an RPN lower than the limit, it can still be addressed if it is the opinion of the team that improvements can and should be made. This failure mode and analysis tool is meant as a process improvement device, and good engineering logic should never be omitted from the process.

Table 5.2 Criteria for Assigning the Severity (SEV) Ranking

Rank	Criteria
1	Unreasonable to expect that the minor nature of this failure will cause any noticeable effect on the product performance. Customer will probably be unable to detect failure.
2 3	Low severity ranking due to minor nature of failure causing only a slight customer annoyance. Customer will probably notice only minor system or product performance degradation. Defects that cause the customer to make minor adjustments to its process but have no other effects might be examples of low severity ranking defects.
4 5 6	Moderate failure that causes some customer dissatisfaction. Customer is made uncomfortable or is annoyed by the failure. For example, increased scrap rate or major process adjustments due to poor processing would be considered a moderate failure.
7	High degree of customer dissatisfaction due to the nature of the failure.
8	Customer will be unable to use the product. The problem results in a shutdown of the customer's processing lines but does not involve safety or environmental issues or noncompliance to federal or state law.
9 10	High severity ranking when a failure mode involves potential safety problems and/or nonconformance to federal or state regulations.

9. All action items should have responsibilities assigned to ensure thorough follow-through, so column 9 would contain the name of the responsible person or party. At this point the efforts of the original QRP team have been completed. The team scribe can document the efforts of the team and submit the information to the quality manager.

10–11. When the action items have been completed, the steps taken and the dates of completion should be documented in columns 10 and 11. The team should then be reconvened to complete columns 12–15.

12–15. New OCCs, SEVs, and DETs should be assigned, and RPNs should be recalculated, based on the new system(s) in place. These will use the same criteria as are used for columns 4–7, cited previously. This then becomes documentation of the continuous process improvement effort. Table 5.3 lists criteria for detection ranking.

Table 5.3 Criteria for Assigning the Detection (DET) Ranking

Rank	Criteria	Probability of Shipping	Method Cm
1 2	Remote likelihood that the product will be shipped containing the defect. The defect is a visually obvious characteristic (e.g., color).	1/16,000 1/1,000	3.25 3.00
3 4	Low likelihood that the product wil lbe shipped containing the defect. The defect is easily detectable with routine analyses that contain a low degree of subjectivity and remote possibility for human error.	1/200 1/100	2.50 2.25
5 6	Moderate likelihood that the product will be shipped containing the defect. The defect is an easily identified characteristic, but the detection method contains some degree of subjectivity or a moderate potential for human error.	1/50 1/20	2.00 1.75
7 8	High likelihood that the product wil lbe shipped containing the defect. The defect is a subtle characteristic that is detectable, but the detection method has a high degree of subjectivity or high potential for human error.	1/10 1/4	1.25 1.00
9 10	High likelihood that the product will be shipped containing the defect. The product is not checked or not checkable. Defect is latent and will not appear at manufacturing location (e.g., defect affects processing of product).	1/2 1/1	0.50 0.25

5.5 QRP Process Flow

The QRP is a chart-based process. The following chart identifies steps in the process. A legend is supplied at the end for the abbreviations used in the chart.

Step	Description	Lead Function	Form	Columns
1.	Identify need.	Management		
2.	Charter a team and obtain membership.	QM		
3.	Identify the customers' requirements.	APP	QDP	1–2
4.	Develop product specification requirements.	APP PROC	QDP	3–4
5.	Identify detection (lab) methods. Document precision.	PROC LAB ANL	QDP	5–6
CHECKPOINT:	Is the method precision less than 30% of the specification range? If not, proceed to step 6. If so, proceed to step 7.			
6.	Develop action plan to reduce method variability, modify the spec limits, or devise a sample plan to accommodate excessive variation.	ANL LAB	QDP	7–10
7.	The quality design plan is now complete. Reiterate the quality attributes, the spec limits, and the sample plan on the quality control plan form to begin the next phase of the QRP process.	PROC PROD	QCP	1–6
8.	Document the corrective actions to be taken for nonconforming quality attributes.	PROC PROD	QCP	7
9.	Define the process parameters that control the quality attributes.	PROC PROD	QCP	8
10.	Define the normal operating limits (operating windows or envelopes) for the process parameters.	PROC PROD	QCP	9
11.	Describe the process parameter control scheme.	PROD	QCP	10
12.	Describe the corrective actions to be taken for nonconforming process parameters (how to react to out-of-control conditions).	PROD	QCP	11
13.	Describe every potential failure of the product (within reason).	TEAM	FMEA	1
14.	Describe the effect(s) of each failure from the customers' viewpoint.	APP	FMEA	2
15.	Brainstorm all potential causes for each failure mode.	TEAM	FMEA	3
16.	Using the information from the control plan, assign an OCC ranking for each documented possible cause.	PROD TEAM	FMEA	4
17.	Using the information in the FMEA, assign a SEV ranking for each documented possible cause.	APP TEAM	FMEA	5
18.	Using the information in the design plan, assign a DET ranking for each documented possible cause.	PROC TEAM	FMEA	6
19.	For each documented possible cause, calculate the risk priority number, RPN. RPN=OCC $\times$ SEV $\times$ DET.	TEAM	FMEA	7
CHECKPOINT:	Is the RPN greater than the limit set by the team? If not, proceed to next CHECKPOINT. If so, proceed to step 20.			
CHECKPOINT:	Are actions needed to reduce the RPN? If so, proceed to step 20. If not, proceed to step 22.			
20.	Develop action plans to reduce the risk priority numbers; assign accountabilities and responsibilities.	TEAM	FMEA	8–9
21.	Document actions taken and then recalculate the OCC, SEV, DET, and RPN values for the new process per step 19.	TEAM	FMEA	10–15

Step	Description	Lead Function	Form	Columns
22.	All potential failures have now achieved acceptable levels of risk. Document the final results of the team effort and publish it to the business team.	TEAM		

Legend	
ANL	Analytical measurements R&D
APP	Applications R&D
QCP	Quality control plan
FMEA	Failure mode and effects analysis
LAB	Plant laboratory
PROC	Process R&D
PROD	Production department
QDP	Quality design plan
QM	Quality manager
TEAM	The entire team

Dibutyl Ether Example

Figure 5.6, Figure 5.7, and Figure 5.8 show examples of what portions of the QRP might look like for a fictional dibutyl ether product. Chart templates can be requested from your instructor to use for practice.

Figure 5.6 Portion of a completed quality design plan (QDP).

Quality design plan for: Dibutyl ether   Page 1 of 1   Date: February 10, 2022

Customer Performance attribute 1	Acceptable limits (If any apply) 2	The quality attributes that control performance 3	Proposed limits for quality attribute 4	Detection method for the quality attribute (lab method) 5	Method capability, cm 6	If method capability is less than 2.50:			
						Recommended actions to improve precision 7	Respon-sible party 8	Actions taken 9	Date completed 10
Cold load	2.4 - 2.8 ft ^{2}	Total polymer	38% ± 4%	IR	4.1	N A			
		Hydroxyl	456 - 556	Wet chemical titration with phenyl whatsis	1.33	Investigate need for an automated method	D R F	Reliable instrument not available at this time	11/2/21
						Devise sample plan to compensate during the short term	J L L	Sample plan implemented at plant lab	2/08/22

Quality Management–Quality Reliability Planning **65**

Figure 5.7 Portion of a completed quality control plan (QCP).

Quality control plan for: Dibutyl ether

Page 1 of 1   Date: February 10, 2022

Quality attributes	Sample plan				Specification limits for quality attribute	Corrective actions for non-conformance (outside limits)	Controlling process parameters	Acceptable limits for process parameters	Process parameter control scheme	Corrective actions for non-conformance (outside limits)
	Sample point	Freq.	Samples No.	Tests No.						
1	2	3	4	5	6	7	8	9	10	11
Total polymer	ST	Each	1	1	38% ± 4%	Isolate batch to repolymerize	Monomer feed rate	Target ± 5 gpm	Mass flow meters	Calibrate the meters
							Reactor temp	345°C ± 10°C	Controlling DMV redundant RTDs controlling the tempered water system	Calibrate the RTDs; check tempered water loop for plugs, etc.
Hydroxyl	ST	Each	2	2	456 - 556	Batch must be scrapped	Hydroxyl of incoming raw material	660 - 700	Vendor COA	Out of spec raw material rejected, sent back to vendor. purchasing notified of the problem
Acid content	ST	Each	1	1	3.5 - 4.0 pH	Isolate batch for acid adjustment with carbonic acid	Acidity of the incoming raw material	2.0-2.5 pH	Vendor COA	Reject, return; notify purchasing
							Reactor temp during cookout	350°C Max	Redundant RTDs controlling the tempered water system	Emergency cooling water added; instrument loop recalibrated

Figure 5.8 Portion of a completed failure mode and effects analysis (FMEA).

FMEA for: Dibutyl ether   Page 1 of 1   Date: February 10, 2022

Give a description of the potential failure	Effect on the customer due to this potential failure	List all possible causes for each potential failure	O C C	S E V	D E T	R P N	Recommended actions for RPNs >125	Respon-sible party	Actions taken	Date completed	O C C	S E V	D E T	R P N
1	2	3	4	5	6	7	8	9	10	11	12	13	14	15
High total polymer	Stiff foam, high cold load	Monomer/base feed radio incorrect;	3	4	2	24								
		Rx temp too high (>360°C)	5	4	2	40								
Low total polymer	Soft foam, low cold load	Monomer/base feed radio incorrect;	3	4	2	24								
		Rx temp too low (<340°C)	4	4	2	32								
Hi or lo hydroxyl	Affects cold load	Contaminated with a diluent; raw material off spec	2	8	2	32	Need to establish a better measuring for incoming raw materials	JJP	Customer supplying quarterly SQC data and plant lab is double checking raw material hydroxyl	2/9/22	4	8	2	64
			4	8	5	160								

Summary

Quality reliability planning (QRP) is an improvement strategy that brings the various functions within a business together to ensure continuity of purpose and understanding, so that product can be made that consistently meets the customers' expectations. There are three components of the QRP: the quality design plan (QDP), the quality control plan (QCP), and the failure mode and effects analysis (FMEA).

The quality design plan (QDP) builds an information base so that all functions of the organization have the same foundational customer and process knowledge, thus greatly increasing the efficiency of the rest of the process.

The quality control plan (QCP) documents the manufacturing process control parameters needed to achieve the desired quality output and provides a path forward for handling nonconforming product should the process fail. Process technicians play a key role in ensuring procedures and practices are properly documented and meet the needs of the business.

The final component is the failure mode and effects analysis (FMEA). The FMEA is a tool employed by safety, environmental, and quality practitioners to evaluate the potential problems you might encounter in the manufacturing process and understand the impact of those problems on the customer.

The quality reliability planning (QRP) is based on charts that outline steps pulling information from the three components (QDP, QCP, and FMEA).

Checking Your Knowledge

1. Define the following key terms:
   a. Failure mode and effects analysis (FMEA)
   b. Quality control plan (QCP)
   c. Quality design plan (QDP)
   d. Quality reliability planning (QRP)
   e. Risk priority number (RPN)

2. A procedure used to document the "baseline" knowledge of the organization in order to improve communication between functions in the organization is the:
   a. Failure mode and effects analysis.
   b. Quality control plan.
   c. Quality reliability planning.
   d. Quality design plan.

3. A procedure used to document the process control schemes used to ensure the manufacture of a quality product is the:
   a. Failure mode and effects analysis.
   b. Quality control plan.
   c. Quality reliability planning.
   d. Quality design plan.

4. A tool used to evaluate potential problems so you can make improvements to prevent them before they happen is the:
   a. Failure mode and effects analysis.
   b. Quality control plan.
   c. Quality reliability planning.
   d. Quality design plan.

5. A procedure designed to ensure proper communications occur between all of the functional groups within a business and establish a system to ensure the manufacture of a product that consistently meets the customers' expectation is:
   a. Failure mode and effects analysis.
   b. Quality control plan.
   c. Quality reliability planning.
   d. Quality design plan.

6. Participants in a QRP exercise should include:
   a. Process technicians.
   b. R&D scientists.
   c. Production engineers.
   d. All of the above.

7. (*True or False*) Process technicians play a key role in the quality control plan (QCP).

8. Which is the first step in the quality reliability planning process?
   a. Failure mode and effects analysis (FMEA)
   b. Quality control plan (QCP)
   c. Quality reliability planning (QRP)
   d. Quality design plan (QDP)

9. (*True or False*) Including operators and technicians in the QDP is a good way to provide training.

10. Process capability is captured during which stage of the QRP process?
    a. Failure mode and effects analysis (FMEA)
    b. Quality control plan (QCP)
    c. Quality reliability planning (QRP)
    d. Quality design plan (QDP)

11. Examining every possible failure mode for your process in order to understand the consequences of process failures from the customers' perspective occurs during which stage of the QRP?
    a. Failure mode and effects analysis (FMEA)
    b. Quality control plan (QCP)
    c. Quality reliability planning (QRP)
    d. Quality design plan (QDP)

12. (*True or False*) Process technicians should not have access to the instruction guide prior to completing the FMEA.

NOTE: Answers to Checking Your Knowledge questions are in the Appendix.

Student Activities

1. *Quality Design Plan (QDP).* You work in a small manufacturing plant. Your product is lemonade. You sell your lemonade in bulk (five-gallon containers) to several food services companies around town as well as directly to consumers by the glass. Using a blank quality design plan template, document customer critical performance attributes. What might be important to your customers? Remember, different customers have different needs and expectations. Establish reasonable spec limits for these attributes. Define some quality attributes that might be used to measure "goodness" relative to your customers' performance attributes. Establish limits for these quality attributes. How might you measure these quality attributes? (Answering these questions should complete columns 1–5 on the quality design plan.) To get you started: Customers are interested in color. They want the color between two certain shades, and they use paint samples to demonstrate goodness. You measure color in your lab with a colorimeter that provides a wavelength output. As long as the wavelength of your measured sample is within specified colorimeter limits, the color seen by the customer should be acceptable.

2. *Quality Control Plan (QCP).* Using three or four of the quality attributes you identified in column 3 of your quality design plan, continue this example by filling in relevant portions of a quality control plan template. In the case of the color that you identified in Activity 1,

specification limits of $\pm 10\%$ from the target have been established. What will you do if you make a batch that is outside this range? (column 7) What is it in your process of making lemonade that controls the color of the final product? (column 8) Establish limits for the process control parameters in column 9. How are these parameters controlled? (column 10) What will you do if the process fails? (column 11)

3. *Failure Mode and Effects Analysis (FMEA).* Take each of the quality attributes from the previous activity and describe a way in which this attribute might fail, using a blank FMEA template. In your example, the color might be too dark, it might be too light, or it might be too cloudy. What might the impact of each of these failures be on the customer? (column 2) What might cause this failure to occur? (column 3) It will be completely subjective, but take a crack at assigning OCC, SEV, and DET rankings for each failure and calculate a risk priority number (RPN) for each using columns 4–7. If any of these potential failures ranks higher than 125, use column 8 to document some recommended corrective actions to improve your process.

4. Provide several examples of QPM. Have the students go through the scoring process, brainstorm on ways to lower the overall RPN scores, improve detection, reduce severity, and reduce likelihood of occurrence.

Chapter 6
Team Skills

"None of us is as smart as all of us."

~ Ken Blanchard

*North American Process Technology Alliance (NAPTA) developed curriculum to ensure that Process Technology courses will produce knowledgeable graduates to become entry-level employees in process technology. Objectives from that curriculum are named here in abbreviated form. For example, (NAPTA Quality, Team Skills 1, Effective Teams 9)" means that this chapter's objective 1 relates to objective 1 of the NAPTA curriculum about team skills and objective 9 on effective teams.

Key Terms

Quality circle—a group of people, typically organized by logical structure within a company, who work together for a common goal, **p. 73**

Synergy—the phenomenon in which the total effect of a whole is greater than the sum of its individual parts, **p. 72**

Team—a small group of people, with complementary skills, committed to a common set of goals and tasks, **p. 72**

6.1 Introduction

Throughout this textbook, you will notice that quality improvement projects are usually accomplished through teams. In this chapter, we discuss what individuals can bring to a team, the purpose for working in teams, and the stages of team building. You will gain insight about how teams work, including stages in the life of a team and the various roles that team members play in order to keep a team on track.

Several quality tools can be applied at different stages in the life of a team (Figure 6.1). Examples include the team charter, an agenda, RACI charts, Gantt charts, and an action item log. This chapter provides explanations and examples of each of these tools to help prepare you for your role as a team member in the process industries.

Figure 6.1 A. Teams consist of two or more people working together toward a common goal. **B.** The team toolbox includes such tools as an agenda, RACI charts, Gantt charts, and an action item log.

CREDIT: A. Photographic Services, Shell International Limited.

A.

B.

Regardless of their role on the team, all members should follow some basic etiquette points to help ensure the team's success. Components of good team etiquette and ways of overcoming challenges that arise are discussed briefly.

Personal Qualities for Effective Team Building

Individuals who come together for a common purpose become a team. In work teams, as in sports, people have different strengths and personalities. Some naturally keep the overall plan and goals in mind. Others will be very effective in accomplishing specific tasks well and within deadlines. Some will be good at communicating everyday updates, such as making sure everyone hears about schedule changes or arrival of new materials.

Being a member of any team requires personal qualities that will be valuable to the team, and more generally to the organization for which you work. Time management, organization, planning, prioritization, and productivity are skills that you probably have developed as a student and will continue to develop as a process technician. In addition, as you start your professional career, you will find that employees who are effective in their jobs and on teams will be patient, task oriented, flexible, adaptable, and confident. They will take initiative and display a strong work ethic. When working with a team, effective employees will acknowledge their own limitations without becoming defensive and will use this information as an opportunity for personal growth. Knowing your own strengths and weaknesses will contribute to your success in many contexts, both professional and personal.

Being a quick learner is a valuable quality, but learning to understand the knowledge and skills necessary to function in an organization or on a team is even more valuable. Employees who show willingness to learn in various environments, accept additional responsibilities, share knowledge and train others, and take ownership of processes outside of their normal duties will be great assets to their organization and to any team on which they serve.

Being a valued employee goes beyond going to work and performing assigned duties. Understanding the company's mission, vision, and values and aligning your own values with the organization's takes you a step beyond. Knowing the organizational structure and your place within it, as well as available resources (i.e., training department, human resources, quality assurance lab, maintenance, and engineering) and how to utilize them will increase your value to your company and provide opportunities for professional growth. These will all increase the effect you will have as a team member.

When working with a team, you will find various personality types. Learning to work together (Figure 6.2) includes the following:

- Getting over the "us" versus "them" mentality
- Being willing to share and participate
- Appreciating diversity
- Valuing others' perspectives
- Displaying resourcefulness
- Aligning individual values and subsequent actions with those of the team
- Recognizing that many points of view are better than one
- Accepting feedback

Figure 6.2 High achievement depends upon a willingness to contribute to the goals of the team.

CREDIT: danleap/Shutterstock.

- Showing a willingness to depend on others
- Appreciating the value of "win/win" thinking
- Understanding team dynamics

Purpose of Teams

Research has shown time and again that employees will take more pride and interest in their jobs and show more ownership in the success of the business if they are allowed to make meaningful contributions to that success by participating in the decision-making processes. The opposite is also true. When management uses teams to justify a decision it has already made, the team concept will fail miserably.

Teams are not a cure-all for quality and productivity problems. Certain issues can be effectively handled by individuals working alone, but oftentimes a team approach is best. Working together in a team allows the organization to achieve results that could never be attained by a collection of individuals, a concept known as **synergy**. Two heads really are better than one.

One college experiment involved four roommates pulling on a rope attached to a tension-measuring device. After each individual was given a chance to pull as hard as he or she could, the roommates were put into groups of two, then three, then all four pulling at one time. The experimenter found that when they worked together, they could pull much more than the sum of each one pulling independently ($1+1+1 > 3$). That is synergy, and that is why teams are such an important part of the quality process.

For a **team** be effective, it is important that it meets the following requirements:

1. **It is a small group.** We can argue back and forth about what constitutes a "small" group of people, but everyone would agree that a group of 5,000 is not a small group and can never operate effectively as a team.

2. **People have complementary skills.** The concept of complementary skills is equally important. Can you imagine a winning baseball team that had eight pitchers, one catcher, and nobody playing the infield or the outfield? Success requires that the team members be able to assist each other and build on each other's strengths. An example of team members assisting each other with complementary skills would be a root cause analysis (RCA). RCA teams generally consist of cross-functional team members from various disciplines.

3. **There are common goals and a sense of commitment.** Each and every person on a team must be committed to the same goal. Sadly, this is not always the case. Too often some team members have their own ideas about what is important.

Consider a typical business unit within the process industry. The safety, health, and environmental director is interested in safety and spill statistics. The manufacturing individuals are interested in making pounds of product. The quality members are working on process improvements. The financial representative is interested only in controlling costs. To function effectively as a team, these players would all have to subject their personal and functional goals to the goals of the team.

Everybody sees work from a different perspective. Let us be honest; most of us would not be working if we did not have to do so. Few of us would continue to work without the benefit of a paycheck. Knowing how other people see the concept of work helps you understand what roles might be best for them in a team environment.

- Some see it as a compulsory activity engaged in solely for the purpose of making a living. These individuals are hard to engage in participatory team activities. They have little motivation to do anything more than the bare minimum required to get by.
- Some see work for its benefits. They work with purpose, knowing their work contributes to the functioning of society and provides for them and their family. These individuals will be excellent contributors in a team environment.

Synergy the phenomenon in which the total effect of a whole is greater than the sum of its individual parts.

Team a small group of people, with complementary skills, committed to a common set of goals and tasks.

- Some see work as a career path. What they do today is a stepping stone to what they want to do sometime in the future. For these individuals, growth is important because it gives them an opportunity to expand their skills and knowledge. These individuals will be excellent contributors in a team environment.

- Others view their work as a status symbol. Their definition of who they are is wrapped up in what they do. For them, work is equated with personal worth. These individuals are often natural leaders as well as contributors in a team environment.

Understanding the skill set and level of commitment of each team member is important when forming a team.

6.2 Types of Teams

Many different types of teams exist in the process industries. The quality function uses teams to accomplish diverse goals. Here are some examples of teams that you may reasonably expect to participate in as a process technician:

- Process improvement teams
- Root cause analysis (RCA) teams
- Six Sigma teams
- Management system audit teams

These are all examples of types of project teams. They are assigned specific tasks to accomplish and are often given a specified time frame in which to complete these tasks. The expectation is that these project teams will disband when the tasks are complete.

There are also standing teams used in the quality function. A standing team, or committee, is a group of people chartered to provide oversight in an ongoing manner. In the quality field, these are often referred to as quality circles. Quality circles are often attributed to Dr. Ishikawa. He formed quality control circles under the sponsorship of the Japanese Union of Scientists and Engineers while an engineering professor at the University of Tokyo in the early 1960s.

In the United States, the concept of the **quality circle** was first used at Lockheed in the 1970s. A quality circle may be made up of all the process technicians of a certain production line. Their job is the never-ending task of monitoring the quality of production and identifying opportunities for improvement. When an opportunity is found, the right answer may be to charter a project team to address the opportunity.

Quality circle   a group of people, typically organized by logical structure within a company, who work together for a common goal.

Another style of team is the ad hoc team. These short-term teams come together with little or no notice to address an immediate need. Here is an example:

The plant has an oversight committee (a standing team) that is charged with monitoring the environmental performance of the plant. When an incident such as a spill occurs, the supervisor on duty may form an ad hoc team to mitigate the impact of the spill as quickly as possible. This team may include representatives from operations, supply chain, maintenance, and even the safety, health, and environment (SHE) organization. When the immediate danger has passed, management may charter a root cause analysis (RCA) team to review the incident to determine why the problem occurred and what improvements may be needed to prevent recurrence. Process technicians would certainly be needed on the ad hoc team and the RCA team. In some companies, process technicians also serve on standing teams such as the environmental oversight committee in this example.

6.3 Stages of Team Development

The techniques employed by teams to achieve their assigned goals depend on the specific purpose of the team: audit techniques for audit teams, RCA techniques for RCA teams, and so on.

Regardless of what type of team you may serve on, the team's process for working together is important for success (Figure 6.3). Teams do not just spring to life fully formed and effective. Instead, they must be cultivated, taught, and given time to mature before they can be effective. Nearly all teams go through a series of developmental stages that are commonly referred to as forming, storming, norming, and performing. At the conclusion of a team's existence, there may also be a final stage, called adjourning, so for the sake of completeness that option is discussed here as well.

Figure 6.3 How a team develops does affect its success.

Forming

Characteristics of the first stage of team development include tentativeness and insecurity. A lot of individuals might be spending as much time trying to figure out what is going on and who everybody else is as they are working on the assigned tasks (Figure 6.4).

Figure 6.4 Forming.

CREDIT: Viktoria Kurpas/Shutterstock.

If you do not know the other team members, there may be a lack of trust. Individuals might be overly cautious about what they say and how they say it for fear of offending. Anxiety about "fitting in" or being able to contribute might make some team members uncomfortable in this stage. Different members of the team may have different opinions about the goals and tasks assigned to the team, about the makeup of the team, or about the approach being used to achieve success.

So how does the team deal with the forming stage? The team sponsor may want to schedule a kickoff meeting that is designed more to help the team get to know each other than to accomplish any tasks. Icebreakers are perceived by many individuals to be silly and even a waste of time, but in reality they can serve a valuable purpose in helping the team get acquainted. Here are two examples of icebreakers that you may choose to use in your teams.

- Paired introductions—each person is to find someone whom he or she does not know and obtain enough information about this person to introduce that person to the rest of the group.
- Scavenger hunt—in larger group settings you can issue a list of traits that each person is supposed to find in the other team members.

Storming

The storming stage of team development is when people are starting to figure out what is required of them and the other team members (Figure 6.5). During this stage, it is normal for conflicts to develop. Some individuals will disagree with the assigned task or perhaps with the techniques the team is choosing to achieve success. For example, because the team is made up of people from different functions or areas of the company, some may be trained in the Five Whys methodology of root cause analysis and others in the Apollo methodology.

Figure 6.5 Storming.
CREDIT: Interior Design/Shutterstock.

It is natural to be uncomfortable with employing a technique with which you are not familiar. There may also be some jockeying for position in the team. There is almost always an assigned leader. In addition to that person, most teams have a natural leader that others just tend to turn to for advice and counsel.

If you have more than one natural leader on the team, expect a little jousting to occur as everybody seeks to find his or her place. In the case of a team that is self-directed, the team may be responsible for electing a leader.

A couple of quality tools may help the team deal with the storming stage. The first is a team charter. A charter is a document that formally outlines the team membership, goals, responsibilities, timelines, and the like, clarifying the expectations of the team. This helps put everybody on an even playing field and removes some of the questions about who should be doing what.

Another useful tool for helping a team through the storming phase may be the use of a RACI chart to document the roles of the various characters on the team as well as outside the team. A RACI chart is a matrix showing who is responsible, accountable, consulted, and informed about team tasks and progress. In the section on quality tools later in this chapter, more information about these tools and an example of each is provided.

Many times, during team discussions, important items are raised that do not directly contribute to resolution of the team's current action items. To allow the team to proceed past these issues, the scribe (recorder) or other team members will add the item to an ongoing list sometimes referred to as a "parking lot." These issues can be recorded in team minutes for communication to team sponsors, quality improvement leaders, or other affected groups for resolution or possible consideration as other team projects.

Norming

When the storms start to pass, the team begins to settle down and establish operating rules that everybody can live with (Figure 6.6). At this stage, the team composition may have changed a little because of the storms, or maybe the same players are there, but they have all discovered their place on the team. This is the first stage in which the team members are quieting down and getting comfortable with the team and their place on the team.

Figure 6.6 Norming.

GROUND RULES FOR MEETINGS   Some teams develop team ground rules that help formalize their agreement to work together. Here are some ground rules that we have seen employed in the past. Your team may want to adopt some of these as mentioned or modify them for your own use. You may find other rules that you need to add to help your team succeed.

- No surprise meetings—all meetings should be scheduled with at least 48 hours of notice so the members can be prepared.

- Agenda published in advance of the meeting—the agenda should include a list of topics, who will lead the discussion for each topic, how much time is allotted for each topic, the expected outcome for each topic, review of old action items and capturing of new action items, and so on.

- Honor time commitments—stay on schedule. It is not reasonable to expect that most team members can stay over if the team does not accomplish its goals in the time allotted.

- Come prepared—if you are assigned an action item or a discussion topic in a meeting, then you should come to the meeting prepared.

- Stay focused—if you have an agenda as described above, then stick to it. Spending a few minutes chitchatting at the beginning of a meeting is okay, as long as it does not interfere with getting the work done.

- Contribute—it is not fair to have some team members doing all the work while others just attend meetings. Everyone on the team has a role to play, and everyone should share a fair share of the workload. In meetings, this also means that everyone should pay attention and voice opinions. Many good ideas are not implemented because a timid team member was afraid to speak.

- Respect each other—team members can have differing opinions; it is perfectly natural. What you cannot allow is divisiveness in the team that prevents the team from working together. All team members must avoid personal behaviors that could prevent the team from working together.

Whatever your purpose in establishing a team, these meeting management techniques will come into play:

- Setting regular or periodic meetings
- Delegating and assigning tasks
- Assessing and allocating resources
- Having methods for managing conflict
- Showing respect for all team members
- Recognizing possible strengths and weaknesses of team members (including yourself).

Performing

Finally, you reach the team development stage that you desire, which is performing (Figure 6.7). The goal of the team leader is to move the team through the various stages of team development as quickly as possible to get it to this spot. At this stage the team is working together to accomplish its assigned tasks. Each team member knows the part he or she has to play and is ready to play that part. There is a strong commitment to the success of the team by each member.

Figure 6.7 Performing.
CREDIT: Rawpixel.com/Shutterstock.

The challenge at this phase of a team's life is to sustain the momentum. Problems and issues will still arise, but by this time the team has an agreed-upon set of rules by which to operate and is in agreement regarding who should do what and by when. To keep the team on track and ensure progress toward the desired goal, many teams employ some type of time chart such as a Gantt chart.

A Gantt chart can be a huge, complicated mechanism for tracking every aspect of a team's performance or something as relatively simple as a timeline for when major milestones must be reached for the team to accomplish its goals on time. As with the other tools mentioned, this will be covered in a little more depth in the section on quality tools.

Adjourning

Teams come to a close for several reasons (Figure 6.8). Sometimes the team is unable to complete its assigned task and is disbanded. Sometimes the project is postponed for reasons completely unrelated to the team itself. Hopefully, most of the teams on which you serve will come to a conclusion because the team successfully completed the assigned task

Figure 6.8 Adjourning.

CREDIT: Viktoria Kurpas/Shutterstock.

and achieved its goal. Whether or not a team has been successful, it is appropriate for the members to be recognized for their contributions. The type and extent of the recognition will depend on how successful the team has been and on the company's policies.

Some people are project oriented. For these people, the successful completion of the team is a reward in and of itself. They look forward to closing out a team's efforts because it means they have finished a task, and this is important to them. Other members may be more people oriented, and they will miss the team interactions. Some of these individuals may actually mourn the adjourning of the team.

Just as a team is kicked off with a special meeting where team members can get to know each other, the team may also close out with a wrap-up meeting where team members can celebrate their success and draw closure to the effort.

6.4 Roles of Team Members

Regardless of what type of team you are on, several roles will exist, either officially or unofficially. The key roles follow:

- The team leader (Figure 6.9)—schedules the meetings; provides progress reports to the sponsor; ensures that the team is making progress according to the schedule; ensures team members are trained to perform within the team (RCA training, audit training, statistics training, and so on); and provides conflict resolution as needed.

Figure 6.9 Team leader

CREDIT: Drazen Zigic/Shutterstock.

- The facilitator—typically a process expert in the team methodology being employed but not necessarily in the technology involved. For example, the facilitator for a root cause analysis team may be an expert in the Kepner-Tregoe methodology but may know nothing about the manufacture of your company's particular product. The facilitator can provide training to the team members and ask questions of the team without having to explore every detail of the issue at hand. The facilitator's job is to make sure the team stays on track. (In some standing teams, the role of facilitator rotates among the members from meeting to meeting.)

- The recorder—somebody has to take notes. Minutes from the meetings provide an organizational memory for the team. In addition to summary notes of what was discussed, the recorder should record decisions made, actions taken, and action items to be completed. Issuing the minutes from meetings provides everyone with not only a record of what they accomplished in the meeting but also a list of what they need to work on to prepare for the next meeting. In many standing teams, the role of recorder rotates among the members.

- The team members—these individuals do the work. They should follow the team's ground rules as well as basic team etiquette: being respectful of their teammates, coming to meetings prepared, arriving on time, actively listening to others, as well as participating themselves, and the like.

Depending on the size and makeup of the team, the facilitator may play the role of timekeeper, or somebody else on the team may play this role. Additionally, team roles may be rotated and may not always be filled by the same person.

6.5 Quality Tools Used by Teams

Team Charter

The charter of a team is the document that spells out the who, what, where, when, how, and why of a team. Formats for team charters vary from company to company, but here are some common elements of nearly all team charters:

- Purpose: Why does this team exist? In broad terms, what is this team supposed to accomplish?

- Specific goals/deliverables: Create in specific and measurable terms and provide details of what must be done for the team to be declared successful. Is it a 10-percent reduction in quality problems? Then define 10 percent of which baseline and specify which quality problem is being reduced. The goal and baseline should be so specific that determining success at the end of the project a no brainer—either the team met its goals or it did not—but there is no room for interpretation.

- Timeline: When is the team supposed to complete the work? Are there milestone dates along the way? Some projects have certain milestone dates to meet along the way. For example, in a typical Six Sigma team there are expected completion dates assigned for each phase (define, measure, analyze, improve, and control) of the project.

- Team membership: Who is on the team and why? When you see a baseball team roster, you see the names of the players and what position they play. Your team charter should include the names of the players and the roles they play on the team as well.

- Sponsorship: Include the name of the person or organization that is sponsoring the team effort. This provides the team with an escalation path should team members run into questions about the purpose or goals of the team or have issues with the team membership.

- Boundaries: It may be helpful for the charter to include boundary conditions for the team. For example, the quality problems that you are working on in the dibutyl ether unit may exclude the new experimental isobutyl ether product line. This helps focus the team on the issue. The team's efforts may be limited to the dibutyl ether production line at plant A but not include work at plants B and C just because of the travel costs involved. In this case, management may desire that the team work on the issue locally at first and then attempt to leverage the learnings to the other plants.

Usually the sponsor will provide a draft charter to the team. Reviewing the charter is a common action for the first couple of team meetings. This gives the team members a chance not only to ensure they understand the charter but also to challenge portions of the charter that may seem unrealistic to them. More than one team has looked at the timeline proposed and told management that what it was being asked for could not be done within that time frame with the resources provided.

RACI Chart

Another commonly used tool to help teams get organized is the RACI chart. This acronym stands for responsible, accountable, consulted, and informed. If a problem requires a team to address it, then it is safe to assume that the problem is not a small, isolated issue. The problem assigned to the team may involve quite a cast of characters, some of whom serve on the team and some of whom do not.

The purpose of the RACI chart is to help the team get a handle on who is affected by the project. Someone has to take responsibility for the work getting done, and somebody is going to be held accountable for everything coming together. Some people (or groups) may need to be consulted along the way. Some individuals just need to be notified or informed of the team's progress. All of these individuals are collectively referred to as stakeholders. They all have a stake in the success of the team, but they all see the team's success from a different perspective.

The odds are that you, in the role of a process technician, will not be asked to complete a RACI chart on your own, but you should know what these charts look like and what purpose they serve. Figure 6.10 (see facing page) is an attempt to identify the roles of the key stakeholders for a family reunion BBQ for each of the tasks involved in that reunion.

Team Ground Rules

The concept of establishing ground rules for the team has already been covered, but for the sake of simplicity, and to make them easier to find later, here they are restated in bullet form:

- Have no surprise meetings.
- Publish agenda in advance of the meeting.
- Honor time commitments—arrive on time and stay on schedule.
- Come prepared.
- Stay focused.
- Allow and encourage everyone to contribute.
- Respect each other.

Remember that these are just a few examples. You will need to work with your team to establish ground rules that make sense for your team.

Agenda

An agenda is a simple, but extremely valuable, tool. For each meeting, the team leader should insist that an agenda be published in advance. This way every team member knows

Figure 6.10 Example of a RACI chart.

RACI chart for family reunion BBQ

Tasks	Parties →	Dad	Mom	Children	Fire dept.	Guests
Develop invitation list		C	AR	C		
Send out invitations		I	A	R		
Notify authorities that dad will be BBQing ...again		R	A		I	
Provide the food for the party		A		R		
Cook the meat		R	A			
Cook everything else			AR			
Attend						AR
Put out the fire		I	I	I	AR	I
Clean up the mess			A	R		

Legend	
R = Responsible
A = Accountable
C = Consulted
I = Informed

what is going to be discussed, why it is going to be discussed, and what preparations are needed to ensure the discussion is value added.

A well-formatted agenda not only helps ensure all the players are prepared to play their roles during the meeting but also helps the facilitator keep the team on track with the timing of the meeting. You may see many different formats used. Figure 6.11 shows an example of how you may design an agenda for your meetings.

Here are the main things that your agenda should include:

- When and where of the meeting
- What topics are to be covered
- Time allotted for each topic
- Desired outcome for each topic
- Who is responsible for leading each topic

If so desired, a spreadsheet can be easily formatted to perform the time calculations so that the user just inputs the duration for each topic and lets the computer calculate the timing for it. When used in this manner, it is easy to have ad hoc or guest members join the team at a certain time for a specific topic of interest.

Action Item Log

An action item log is a simple listing of what the team members agreed to do, when they agreed to have it done, who would be responsible for getting it done, and tracking of their

Figure 6.11 Example of an agenda.

Quality project team meeting

Date:	13-Jun-21
Call-in number:	1.800.567.5349
Facilitator:	Jim
Recorder:	Kim

Timing (CST)		Agenda item	Person(s) responsible	Desired outcome	Comments
Start	Duration				
8:30AM	0:15	Time out for safety	(All)	Share knowledge	
8:45AM	0:05	Agenda review	Jim	Review meeting timeline and objectives	Identify participant time conflicts Identify facilitator / action recorder for next meeting
8:50AM	0:20	Action item update	Angelica	Update action items from previous meeting	
9:10AM	0:30	Review financial performance	Scott	Identify new actions to meet teams goals	
9:40AM	0:30	Review project timeline status (Gantt chart)	Jack	Ensure team on track to meet the timing goals set down by management	
10:10AM	0:15	Update on engineering cost reduction	Cindy	Share project status with team	
10:25AM	0:10	Break	All	Relief	
10:35AM	0:45	Review customer data	Chris	Need a decision from the team regarding how to use the customer survey data. Three options to be presented	
11:20AM	0:40	Brainstorming exercise	Charlie	We need more ideas about how to improve the logistical performance of our operation	
12:00PM		Adjourn	Jim		

performance in getting it done. Figure 6.12 is a portion of an action item log that highlights actions that are not completed and actions that took longer than expected.

Gantt Chart

The last tool to include in the team toolbox is the Gantt chart. A Gantt chart shows all the various tasks that a team needs to complete and when they need to be completed. Usually this is done in a graphical format that makes it easy to see the progress of the team at a glance. Do not expect to be asked to put together a Gantt chart. The reason for including it in this text is to provide some description of the format and to illustrate what purpose this tool can play in the life of a team.

In Figure 6.13, many of the tasks must be done in a sequential order because they rely on the results of previous tasks. Some of the tasks (e.g., Major Task I) have subtasks that must be completed in order to accomplish the major task. Some Gantt charts use color coding to make it easy to distinguish confirmed tasks versus tentative tasks, or maybe to highlight the different functions within the organization so that managers can see at a glance who is responsible for what.

Time Management

Gantt charts and other organizational tools are useful, but they will be pointless if team members do not know how to practice good time management. Personal time management is a process of organizing life events in a way that allows a person to plan, to break large jobs down into smaller tasks, and to balance different areas of life. It involves making a map of

Figure 6.12 Example of an action item log.

Dibutyl ether quality improvement team action log

Assigned	Description	Lead	Due date	Status/action taken	Complete date
1/5/21	Set up a control chart for results on dibutyl ether production line	Charlie	3/1/21	Complete	2/28/21
1/5/21	Contact 5 customers (at random) to conduct mini-survey regarding recent shipment troubles	Angelica	1/13/21	Contacted Ultimate Futility Inc; BooBooButyls; ACME Chemicals-R-Us; I-Glow Chem, and Mega Futile Futures. Results tabulated for presentation at next team meeting	1/14/21
1/5/21	Finalize 2021 financial goals and publish to team members	Scott	2/1/21	Complete	2/1/21
2/10/21	Determine if we should attend quality convention in Orlando	Jack	2/15/21	We think we should go to the convention in Orlando; management is reviewing	2/15/21
2/10/21	Prioritize customers for full-blown customer survey in last half of year	Chris	3/1/21		
2/10/21	Set up meeting with engineering department regarding improvement projects	Cindy	2/25/21		

Figure 6.13 Example of a Gantt chart.

Project title: Quality team

Today is November 3, 2020

Description of the task(s) to be completed:	Team leader	Start date	End date	Days	Confirmed/ tentative	Status
Tasks A	Angelica	1/1/21	2/10/21	40	C	
Tasks B	Jim	1/1/21	4/1/21	90	C	
Tasks C	Jack	2/1/21	3/1/21	28	C	
Tasks D	Cindy	2/15/21	4/1/21	45	C	
Tasks E	Chris	3/1/21	4/1/21	31	C	
Tasks F	Charlie	3/1/21	10/28/21	241	C	
Tasks G	Melissa	3/15/21	12/30/21	290	T	
Tasks H	Jerry	1/3/21	1/22/21	19	C	
Major task I	Scott	2/1/21	3/15/21		C	
Minor task I-1	Liz	2/1/21	2/20/21		C	
Minor task I-2	Dylan	2/15/21	3/1/21		C	
Minor task I-3	Josh	2/28/21	3/15/21		C	
Tasks J	Crystal	4/1/21	5/1/21	30	C	
Tasks K	Chloe	1/6/21	1/22/21	16	C	Cancelled
Tasks L	Adrian	3/16/21	4/1/21	16	T	

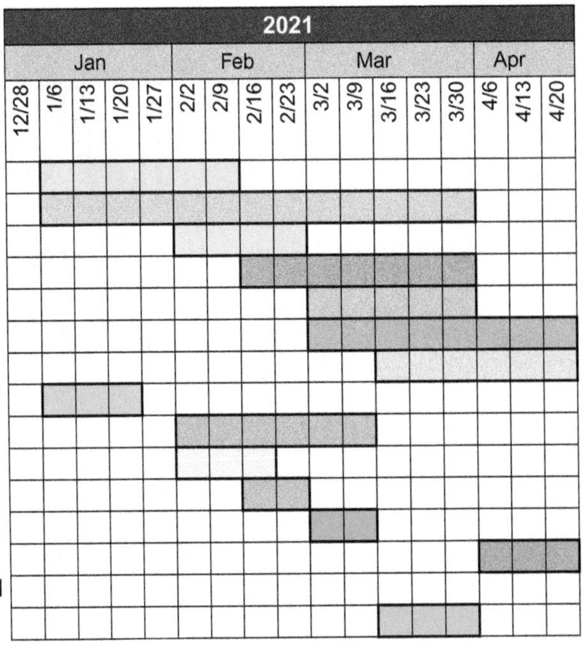

the major pieces of life and seeing how those pieces fit together or can be made to fit together better. Some common elements in time management include the following:

- Listing basic functions that a person performs on a regular basis (sleeping, eating, weekly repeated activities such as church or gym)
- Identifying significant dates such as birthdays, dates of child's graduation, and so on
- Blocking out work times and vacation times on a calendar
- Identifying priority deadlines for work
- Regularly reviewing the "path" to deadlines in terms of work hours and steps to be taken
- Assessing progress on specific steps against the overall plan, and reshaping the steps or plan as needed to reach the goal

6.6 Team Etiquette

Working together as a team does not always come easily. It can be made easier if everybody on the team practices team etiquette (Figure 6.14). Here are a few tips that support success in team efforts:

- Listen carefully—if you are not listening to others, you will not know what is occurring and will not be prepared to participate in the discussion.

Figure 6.14 Team etiquette.

It is okay to disagree as long as you aren't disagreeable!

- Watch carefully—pay attention to nonverbal clues as well as to what people are saying. Have you ever noticed somebody sitting back in a team meeting with his or her arms crossed and eyebrows furrowed? Sometimes what people do *not* say speaks volumes. Pay attention to these nonverbal forms of communication as well. Get the issues out on the table so everybody can participate.

- Participate actively—teams succeed when each individual on the team fulfils his or her role. Do not be an absentee member even when you show up. Speak your mind and make your opinion known.

- Contribute clearly—sometimes your message will come across clearly; sometimes it will not. Each member on the team comes to the table with a different set of experiences and biases. How you say what you want to say is just as important as the words you use. Think about the message you are trying to communicate. Think about the best way to get that message across. Decide whether you need a drawing or an example to illustrate your message. Voice your thoughts clearly. If your message cannot be heard in the back of the room, it cannot be understood.

- Respect others—just as you would have them listen to you and value your opinion, you should listen to them and value their opinions. Avoid sarcasm, put-downs, name-calling, and so on. These communication techniques never build consensus.

- Focus on the job at hand—if possible, have all team members turn off their cell phones and spend team time working on team issues. Multitasking is almost always assumed to be a sign of a productive person, but in many cases attempting to work on too many things at once really results in doing none of these tasks well. Take the time to focus on the team effort.

Factors That Negatively Affect Productivity

Many factors can affect a team's productivity, and every team member can increase or decrease the effectiveness of the team. The points will lead to team success. The following are a few factors that can limit a team's success.

- Being late. Making others wait naturally causes loss of work time. It also lowers morale. Consistently being late sends the message that you (and your time) are more important than anyone else in the room.

- Paying poor attention. Having side conversations, taking phone calls or texting, and simply losing focus on the topic are all actions that have a negative impact on the team.

- Doing the minimum to get by. It is very easy to see the difference between workers who want to leave things ready for the next person and those do the minimum work required before clocking out. The first group sets the stage for the process to continue smoothly from one shift to the next. The second group reaches the bare minimum of what is required. In the long run, the behavior of the second group can lead to missed deadlines, production of off-spec product, and potential for creating hazards for the team.

- Having a negative attitude. Employees have a choice each day to approach work as a challenge or as a trial. Those who approach it as a challenge are in a better frame of mind to learn and to work with others. They will get satisfaction from the progress they make. Those who face work with a list of complaints not only close off their own creativity; they also create a toxic work environment for others.

6.7 Conflict Resolution

Entire books are written on the subject of conflict resolution, and you could easily fill a book with the topic. For the purposes of this chapter about working in teams, here are just a few points about conflict and conflict resolution for you to consider:

- Expect that conflicts will arise on your team. Get used to the idea. Do not let it bother you. Just be prepared for it.

- Avoid making conflicts personal. Focus on the concepts, not the people. It is okay to disagree, as long as you are not disagreeable.

- Seek solutions that everybody can accept (Figure 6.15). I think "A-42" is the right answer. You think "B-42" is the right answer. Neither of us can agree with the other's option. A good conflict resolution technique for the leader to employ would be to look for common components in our opinions. Because we both agree that the solution should include "-42," what compromise would take advantage of this common link and allow us to move forward? Perhaps there is an "AB-42" option that would make us both happy.

Figure 6.15 Five-step model for conflict resolution.

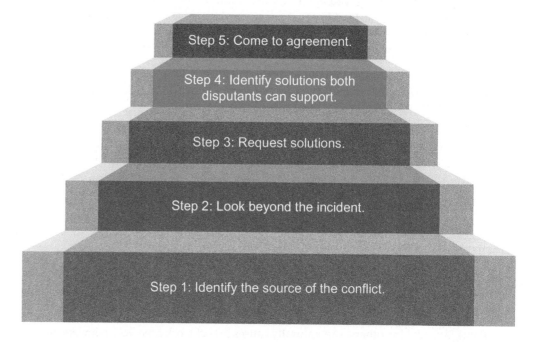

Step 5: Come to agreement.

Step 4: Identify solutions both disputants can support.

Step 3: Request solutions.

Step 2: Look beyond the incident.

Step 1: Identify the source of the conflict.

- Deal with the conflicts as they arise. Do not let them sit unresolved in the background. They will affect the workings of the group and will only get worse with time.
- Recognize that sometimes the conflict you see in the team has its origins outside of the team. Sometimes individuals on a team simply do not get along. Team members do not need to be best friends. They just need to agree to work together toward the team goal.

Summary

Teams are an important means for getting work done in the process industries, and everybody can expect to serve on a team of some kind, in some role, eventually. Many of the quality improvement processes discussed in this book rely heavily on teamwork in order to achieve the desired goals. If you look back to Chapter 1, *Introduction to Process Quality*, you will be reminded that Dr. Juran taught about the importance of using breakthrough teams to accomplish quality improvement.

By working together in effective teams, we can accomplish more as a group than the sum of what we can accomplish individually; this is the concept of synergy. In the realm of quality, we use the team approach for audits, Six Sigma projects, root cause analyses, and other objectives.

Regardless of what kind of team we are on, all teams go through a series of stages of development, which include forming, storming, norming, performing, and adjourning. Recognizing that these stages are a natural evolution of the team helps us understand what the team is going through.

You can employ several team tools such as an agenda, ground rules, Gantt charts, action item logs, and RACI charts to help the team get through the more difficult stages of development and become effective more quickly.

There are many roles to be played on each team, and during different parts of your career you may well play them all at one time or another. Roles include team leader, facilitator, recorder, and team member. No matter what role you play, you can always make the team a better place to be by following simple team etiquette guidelines, most of which stem from the golden rule: do to others what you would want them to do to you.

Checking Your Knowledge

1. Define the following key terms:
   a. Quality circle
   b. Synergy
   c. Team

2. (*True or False*) Synergy is the idea that teams working together can accomplish more than the sum of what can be accomplish individually.

3. Which of the following qualities are useful soft skills to develop for working on a team? (Select all that apply.)
   a. Patience
   b. Flexibility
   c. Sophistication
   d. Confidence
   e. Adaptability

4. Which of these answers is NOT a part of this chapter's definition of a team?
   a. Small group
   b. Highly trained participants
   c. Complementary skills
   d. Common goals

5. What types of quality teams might you serve on as a process technician?
   a. Audit teams
   b. RCA teams
   c. Six Sigma teams
   d. All of these and more

6. The stage of team development that is characterized by tentativeness and insecurity is:
   a. forming.
   b. storming.
   c. norming.
   d. performing.

7. The stage you want to get your team to as quickly as possible, because this is where the team members are getting the job done, is called:
   a. forming.
   b. storming.
   c. norming.
   d. performing.

8. In the early life of a team, there is often some jockeying for position and infighting among the members. This stage of team development is called:
   a. forming.
   b. storming.
   c. norming.
   d. performing.

9. Which of these roles might process technicians find themselves assigned to in a team?
   a. Team leader
   b. Facilitator
   c. Recorder
   d. Team member
   e. Any of the above

10. (*True or False*) A team charter is a tool that shows a graphical timeline of the activities that need to be accomplished.

11. (*True or False*) A RACI chart is used to document the roles of the various stakeholders in a project.

12. (*True or False*) A really good worker can get a lot more done alone own than on a team.

13. (*True or False*) Dr. Joseph Juran was a firm believer in the team approach to quality improvement.

14. (*True or False*) Setting ground rules for your team is an excellent way to establish order and help your team work together effectively.

15. (*True or False*) If your team leader is good at the job, there will not be any conflicts on the team.

16. (*True or False*) Treating others with respect—as you would have them treat you—helps make the team a more pleasant working environment.

17. (*True or False*) Team members with less experience than you will benefit if you act confidently, even if you do not know the answer to a question.

NOTE: Answers to Checking Your Knowledge questions are in the Appendix.

Student Activities

1. Teams are not employed just at work. Many of us serve on committees outside of work. If you participate in any kind of a team, draft a team agenda for your next meeting. Did you find that the meeting went more smoothly when it was planned?

2. Have students share a time when they were part of a team that struggled. What were some of the negative outcomes? Ask them to identify one factor in the failure of the team that they think is important for the success of most teams. Brainstorm what could have been done differently to improve the outcome of the team's efforts.

argon
column
(low ratio)

Chapter 7

Continuous Improvements—Root Cause Analysis (RCA) and Corrective Action/ Preventive Action (CPA)

"Erroneous assumptions can be disastrous."

~ PETER DRUCKER

Objectives

Upon completion of this chapter you will be able to:

7.1 Explain the purpose of root cause analysis. (NAPTA Quality, Data Collection 9*) p. 90

7.2 Explain the processes technician's role in root cause analysis. (NAPTA Quality, Data Collection 9) p. 91

7.3 Compare and contrast the most commonly applied root cause analysis methodologies (i.e., Kepner-Tregoe, Five Whys, and Apollo). (NAPTA Quality, Data Collection 9) p. 92

7.4 Recognize and identify cause-and-effect diagrams, interrelationship digraphs, and current reality trees (NAPTA Continuous Improvement and Corrective/Preventive Action 8). p. 97

*North American Process Technology Alliance (NAPTA) developed curriculum to ensure that Process Technology courses will produce knowledgeable graduates to become entry-level employees in process technology. Objectives from that curriculum are named here in abbreviated form. For example, "(NAPTA Quality, Data Collection 9)" means that this chapter's objective 1 relates to objective 9 of the NAPTA curriculum about data collection and control charts.

Key Term

Root cause analysis (RCA)—the analysis of data to determine the true root cause of a problem which, when identified and corrected, eliminates a problem forever. Also called *root cause investigation*, **p. 90**

7.1 Introduction

This chapter focuses on a quality topic that is more procedural: root cause analysis (Figure 7.1). As the name implies, root cause analysis means examining the information to determine the true underlying cause of a problem or incident. This means digging deeper, not just pointing to the first contributing factor.

Figure 7.1 Root cause analysis (RCA).

CREDIT: rolandtopor/Shutterstock.

When a problem occurs, it is almost shocking how often people jump to conclusions about what caused it, completely ignoring any real analysis of the data or events surrounding the situation. Yet, in order to improve processes, it is essential to eliminate the root cause, not just the contributing causes or factors related to the problem. (There will be further discussion of the importance of eliminating root causes in Chapter 8, *Continuous Improvement— Six Sigma*.)

This chapter presents several different styles of root cause analysis, such as Kepner-Tregoe, Five Whys, and Apollo. It also outlines the role of the process technician in supporting these efforts. There are a few tools commonly used in root cause analyses, or RCA, that you will soon be able to recognize by sight. One such tool is the cause-and-effect diagram, described briefly here and also presented in Chapter 11, *Other Basic Quality Tools*. In addition, this chapter introduces a couple other new tools: interrelationship digraphs and current reality trees.

A fact that has been mentioned and will be covered again in later chapters is that what is important is getting the job done, not applying one particular tool instead of another. As long as the methodology gets to the root cause, it is worthwhile. There are lots of tools in the quality toolbox. It is just a matter of choosing one that is effective and will get the job done.

Purpose of Root Cause Analysis

Root cause analysis (RCA) the analysis of data to determine the true root cause of a problem which, when identified and corrected, eliminates a problem forever. Also called *root cause investigation*.

When something unplanned, unexpected, or undesired happens in a process, it is important to identify the true root cause in order to eliminate it and prevent future recurrence. The method used will depend on the company you work for and its preferred methodology. Failure to conduct a thorough **root cause analysis (RCA)** may end up costing the company a lot of time, money, and resources fixing the wrong thing, and then having to find and fix the right thing.

Think about your car or truck. It is making a rattling noise that is driving you crazy. You spend half of your day in the garage working on the problem and discover the area where the rattle is coming from, so you make a temporary fix to stop the rattling noise.

Unfortunately, you addressed the symptoms but not the root cause. Perhaps a bolt or bracket has vibrated loose and is allowing a hose to swing back and forth and make noise. You find the loose bracket and replace it. Is this good enough? What caused the bracket to come loose in the first place? Maybe it was a poor design. Maybe your car has over 150,000 miles on it and the bracket came loose from normal wear and tear. But perhaps there is a deeper cause. Maybe the engine is vibrating more than it used to because of a shaft imbalance. Perhaps the flywheel has lost a tooth and is no longer balanced.

Sometimes a problem is obvious, but oftentimes a more thorough investigation into the problem is needed to determine the true root cause. How deep do you have to dig? It is rare to complete a root cause investigation with a couple of people in an hour or two.

The purpose of root cause analyses is to identify the underlying reason(s) behind a problem. There may be more than one root cause. There may be a combination of root causes with mitigating circumstances. Things that do not cause a problem in the winter may cause a problem in the heat of the summer. All the relevant data must be evaluated. It is by sorting through all the data that you will find the combination of causes and circumstances that allowed your problem to manifest itself. Then, and only then, can you apply corrective action to ensure that this problem will not occur again.

7.2　Role of the Process Technician in Root Cause Analysis

As a process technician, you are on the front lines of the process industries. You are there where the action is around the clock. Process technicians will be involved in any root cause analyses that involve the operation (Figure 7.2). Modern plants have distributed control systems (DCSs) that record thousands of pieces of information every hour. These computerized systems hold a wealth of raw data but are oftentimes devoid of usable information. They will never replace a human being's ability to see what is happening around the plant and establish relationships between those events and the readings on the computer.

Figure 7.2　A. and **B.** Process technicians involved in reviewing and recording data.
CREDIT: A. Gorodenkoff/Shutterstock; **B.** Fusionstudio/Shutterstock.

A.

B.

The discussion has already mentioned that the scope of a root cause analysis may be broad or narrow, may consume many weeks or just a few hours, and may involve many people or just a few. An informal analysis is often just one person talking to a few others to document the issue.

A formal root cause analysis typically involves a team of people assigned the role of understanding an issue in depth. This team should include the individuals involved in the issue as well as individuals that were not involved. It should have a mix of technical experts as well as hands-on experts. The makeup of the team will depend on the nature and complexity of the issue.

As a process technician, you can expect to participate on many of these teams during your career, sometimes as an ad hoc member called in to answer just a few questions, sometimes as a full-fledged team member. However you are asked to participate, it is in your best interests to help the team put all the pieces together so the right issue gets fixed the right way to prevent recurrence.

7.3 Root Cause Methodologies
Kepner-Tregoe Methodology

(For more information, visit **www.kepner-tregoe.com**.)

Dr. Charles Kepner and Dr. Benjamin Tregoe were researchers employed by the Rand Corporation in the 1950s, conducting research in decision-making breakdowns at the Strategic Air Command. Finding that rank was less critical to successful decision making than were rational thinking and the use of logical processes, they developed a methodology to guide individuals through a logical thought process for gathering, organizing, and analyzing information before making decisions.

The Kepner-Tregoe company is still going strong and still supporting industry, having celebrated 60 years of success in 2018. Obviously, this chapter cannot cover in depth what Kepner and Tregoe cover in an entire book, but here is a brief overview of how they approach root cause analyses.

In the methodology that Kepner-Tregoe associates employ, the root cause of a problem is called a deviation. As with most processes, their process is broken down into several steps. The first step is to outline the details of the deviation. To define the problem, the root cause analysis team answers the typical questions, such as who, what, when, and where.

The unique part of the Kepner-Tregoe (K-T) approach is that it also asks the negative of each question. Who was *not* involved? Where did the problem *not* occur? What did *not* happen? Their approach is to document the "is" versus the "is not." Table 7.1 outlines the types of questions that may be asked in the problem definition step.

Table 7.1 Sample Questions in the Problem Definition Step

	Is	Is Not
What	What is the defect? Be specific. Name the product or the equipment or the business unit.	What is not defective? Again, be specific. What products or equipment or businesses were not affected?
Who	Can you name a specific person or shift or contractor company that was involved?	Who (person, shift, company) was not involved?
Where	Where did the problem manifest itself? Where was the problem seen? Was the defect on a certain location on the product or in a certain geographic location?	Where was the problem not occurring? Where was it not seen? What other places might you have expected to see the problem but did not?
When	When did the problem start? When did it stop? Did the timing coincide with another event such as a shift change or a change in the raw materials?	When was the problem not a problem? When else might the problem have been seen but was not?
Severity	How bad was the problem? How many units of product were affected? How many dollars did it cost?	How many other units might have been affected but were not? How severe might the problem become if left unattended?

The next step in the K-T process is to develop hypotheses as to the possible cause(s) of the deviation. For each hypothesis, the K-T process looks for distinctions and changes that might give clues about the root cause. A possible root cause that is identified might be based on the differences between the "is" and "is not" documentation generated in the first step. This process uses the data from the first step to generate a small list of potential causes that you have reason to suspect.

The third step in the K-T process is to test these hypotheses to see whether they logically explain every "is" and "is not" that is listed. This narrows down the amount of work that has to be accomplished in the final step of the process by eliminating potential root causes logically without the need for data collection.

The fourth and final step in the K-T process is to validate the potential causes as root causes by collecting data. The team will have to decide what is the easiest, cheapest, and quickest way to collect these data. This has to be done because without validation, you cannot be sure you are fixing the right item. Table 7.2 shows a K-T plan that clarifies how this methodology is used to solve problems.

Table 7.2 Example of a Completed Kepner-Tregoe (K-T) Process

Deviation Statement: Three bad batches of product were produced in the plant in September.

	Is	Is Not
What	Three batches of dibutyl ether failed the water specification.	All other specification parameters were normal.
Who	The IBF Line II production team.	A specific shift, as these batches were made over the course of several days with several shifts involved.
Where	Problem was with dibutyl ether manufactured in the IBF Line II facility.	Affecting IBF Line I.
When	The batches were the last three in the campaign. Starting September 12 and no end date. The plant shut down to prevent further off spec product.	A problem in August or at the beginning of the campaign, which started on September 2.
Severity	Three batches failed and the plant shut down. In addition to discarding these batches, the company is losing money by not running.	All products produced in the plant.

With this information in hand, the root cause analysis team noticed that the company had switched to a new lot of a certain raw material on the same day that the problem began. Team members then developed the hypothesis that the problem was caused by a bad batch (lot number X15) of this raw material. They noticed that lot X15 of the material was used in the manufacture of all three of these batches and was not used on Line I at all. This lot of raw material was not staged for use until September 12, when the previous lot ran out on Line II.

This result was common to all of the shifts. It appeared to meet all of the criteria of the "is" and "is not" definitions, passing the company's logical review. Every other theory that the team came up with failed to satisfy the "is"/"is not" criteria. The last thing for the team to do was to validate or invalidate this potential cause with data.

One obvious way to proceed would be to make another batch of dibutyl ether with raw material X15 and a batch with lot number X14, which was discovered to have made good product prior to the problem. Management might have a problem with this validation method, though. If the team's hypothesis is right, the company would make yet another bad batch of product that would cut even further into its profits.

Another way to proceed would be to send samples of lots X14 and X15 to the laboratory for testing to see whether something is different about them. A few lab tests could be quicker and cheaper than making commercial-sized batches of product.

The team might even contact its supplier of the raw material and ask whether it saw any differences in the batches. Enlisting the support of suppliers is an excellent way of understanding the process and helping them to understand it as well. It may be that the supplier noticed a difference but did not think it would be important. When in doubt, ask. It is easy to undercommunicate but almost impossible to overcommunicate.

Five Whys Methodology

Let us now look at one of the most commonly used root cause analysis techniques applied in the process industries. It is a simple application of root cause analysis called the Five Whys (Figure 7.3). As always, it starts with a problem statement. The root cause analysis team asks the question, "Why did this occur?" and documents the response. It may be that data are needed to answer the question. The response to the first why question is now treated as the new problem statement, and the team asks the question, "Why did this occur?" This process is repeated five times to determine the root cause of the original problem statement.

Figure 7.3 Five Whys.

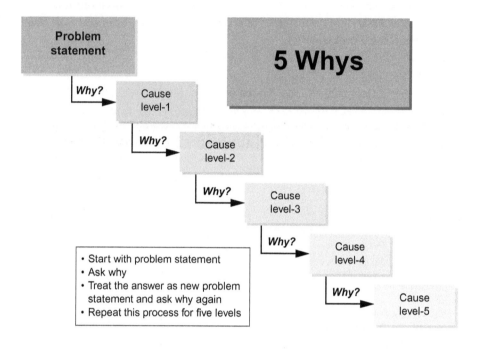

Let us take a look at an example of how this might work itself out in a plant (Figure 7.4). A department tracks product quality on every batch it makes. During the month of September, it had a bad month and made three bad batches of product that failed the specifications and could not be sold to customers.

Figure 7.4 Practical example of Five Whys.

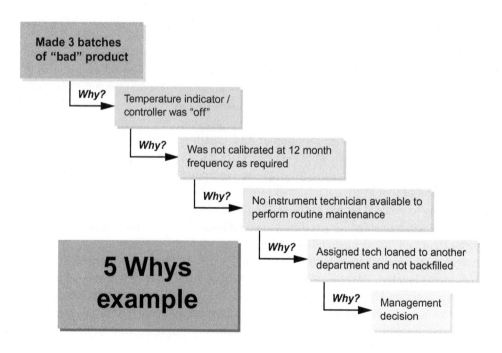

The cost of making this bad product must now be absorbed out of profits. The boss is not happy and charters a root cause analysis team. The team documents the problem statement and conducts a brainstorming exercise. Somebody mentions that the particular failure the department encountered indicates a temperature problem in the plant. You check it out and, sure enough, the temperature indicator is not reading correctly.

The department thought it was running at 212 degrees Fahrenheit (100 degrees Celsius), but it was really running at 230 degrees Fahrenheit (110 degrees Celsius). This difference is enough to explain why the three batches of bad product occurred. In some companies the team might stop here. It has found the root cause and fixed it.

However, stopping means missing an opportunity to prevent recurrence. Instead, this team's members treat the temperature controller issue as the problem statement and asks themselves, "Why was the temperature controller off?" With a little investigation, they find out that the device was scheduled for routine calibration every 12 months but has not been calibrated for more than 18 months.

Again, this process digs deeper, asking, "Why was the device not calibrated on schedule?" It is discovered that there is no longer an assigned instrument technician.

"Why not?" It turns out that the technician was loaned out to another department, and no backup was identified. Notice that this statement is really two answers. Reassigning the instrument technician to another department for a special project or shutdown is a fairly common occurrence. The part that is negative is that no provisions were made to fill that position in the meantime.

The team asks, "Why not?" one more time. Management admits with embarrassment that it just made the decision and accepted the risk.

A couple of things to notice:

- You may have to ask why four, five, or six times, but eventually you will get to a real root cause. The number of levels is not fixed at five. That is just a guideline to press team members to take questions deeper than they may otherwise do.

- You need to have data to make decisions. Intuition is okay as a guide, but data validate what is actually happening. Sometimes this means the RCA team has to get together and generate a list of desired information, then dismiss for a day or two while the data are gathered, then come back together to review the information and plan next steps. Sometimes there is more than one answer to the question why. In this case, the diagram will get more complicated as each one of the new branches is explored further. This is normal. Many problems you encounter will not be solved with a few quick questions.

Apollo Methodology

(For more information, visit the Apollo home page.)

Another widely implemented approach to root cause analysis is the Apollo root cause analysis methodology. The Apollo root cause methodology is well ingrained in process industries, with clients such as Hercules, Valero, Huntsman, Rohm and Haas, and Dow Chemical.

Again, it is difficult to cover the detail that the Apollo methodology deserves in just a few pages of this text. This section gives an overview so that you will be familiar with the basic approach when you are ready to take your place in industry and participate in root cause analysis.

The Apollo root cause analysis methodology is similar to the Five Whys approach in that the answer to each why question becomes the new problem statement. The team repeats the process to keep digging deeper and deeper into the issue. In contrast to the 5 Whys, in the Apollo methodology, the team does not ask why. Instead, they ask "Caused by?" The answer to each "Caused by?" question is not a single, simple response. The Apollo method looks at the condition (or conditions), the action (or actions), or a combination of all of these ("accepted practice," procedure use, and so on) that exist to cause the problem to occur.

The beauty of the Five Whys methodology is its simplicity. In contrast, the beauty of the Apollo approach is a recognition that most problems are not that simple. There may be more than one action and more than one condition that merge to allow a problem to occur. Many times, interconnected actions and conditions come together to cause a problem to exist.

Consider the example of the employee who got a slipped disc, illustrated in Figure 7.5. The action that the employee was taking was walking on the unit floor. The floor was found to be wet and slippery. The employee was also lifting a box of supplies from the supply room floor. You have at least two actions (walking and lifting) and two conditions (wet floor and slippery floor) to consider.

Figure 7.5 Example of an Apollo style cause-and-effect diagram.

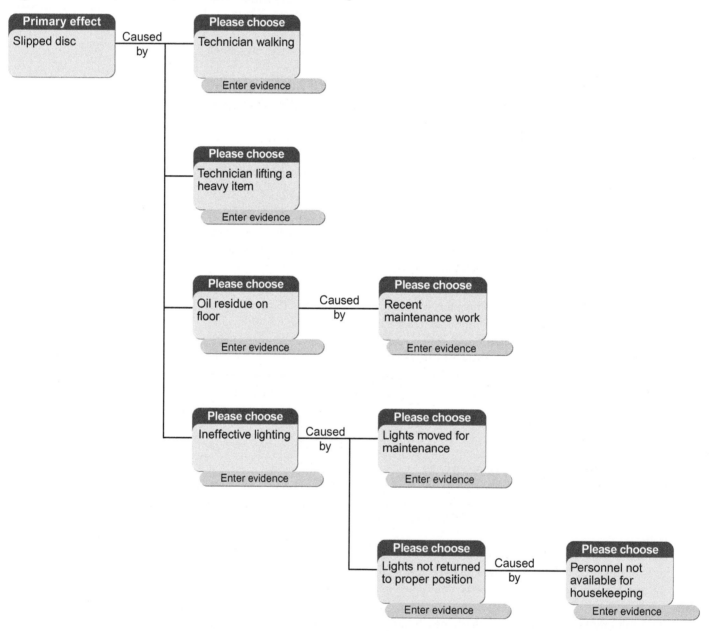

Each one of these four causes must now be examined as if it were the problem statement. What caused the employee to be walking in the unit? What caused the floor to be slippery? What caused the room to be badly lit? What caused the employee to be lifting a box of supplies?

For each of these problems, seek to identify the actions taken and the conditions that existed. The room was poorly lit because lights had been moved and had not been replaced yet. What caused them not to be replaced? Lack of manpower? An unreported problem? The floor was slippery because of the lack of proper housekeeping after maintenance work. No "Wet Floor" sign had been posted, and no communication had been issued. What caused that failure?

As you can see, the cause-and-effect diagram for something as simple as an employee's slipping on a floor can be fairly large. In the end, you do not identify a single root cause, not even a single action with a single condition. Note that there were several actions taken and that several conditions existed which all contributed to the problem. Mitigating one or more of the conditions and/or actions might have been enough to prevent the employee's injured back.

7.4 Additional Tools for Root Cause Analysis
Cause-and-Effect Diagrams

Cause-and-effect diagrams are also called fishbone diagrams or Ishikawa diagrams (Figure 7.6). With this technique, you start with a problem statement and then brainstorm potential causes. Each cause is then placed on the diagram in the most appropriate category. Commonly used categories are manpower, methods, machines, and materials.

Figure 7.6 Example of a fishbone diagram.

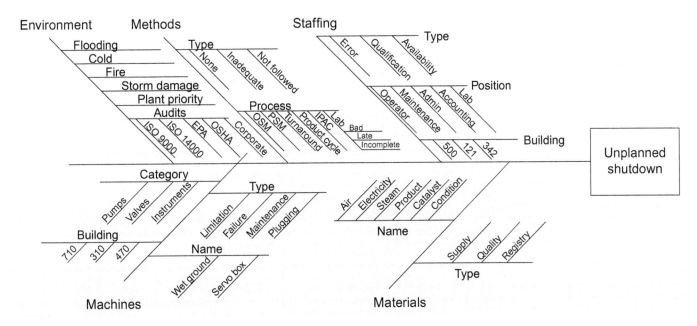

At this point, you have a list of potential root causes, not a root cause. If all you do is generate diagrams of all possible causes, you end up with diagrams on the wall and no idea how to fix the problem. From this list of potential root causes, you must gather data to either validate or eliminate the potential cause. This is obviously more tedious than just generating the diagram. Certain organizations gather data on all identified potential root causes, but more commonly individuals narrow the list down to the probable root causes and then gather data on only those few things.

Interrelationship Digraphs

Another RCA technique that is applied in the process industries is the interrelationship digraph (also called the relations diagram). To use this tool, you basically conduct the same type of brainstorming exercise that is used to come up with contributing factors and plot them all on a chart. Draw lines from any box that might have contributed to the other boxes, and add up the number of inputs and outputs seen.

The result is a diagram that shows how each of the pieces of this puzzle might be related to the other pieces. The root cause that needs to be addressed is the box that has the most influence on the other boxes. Figure 7.7 shows an interrelationship digraph for the employee who slipped in the supply room and broke his leg.

Figure 7.7 Example of an interrelationship digraph.

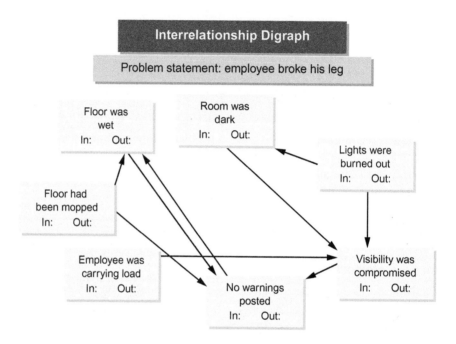

This tool is less often employed in the manufacturing departments. More often it is used by groups of people who are perhaps less technically oriented, such as human resources departments, but you should recognize an interrelationship digraph when you see one.

Current Reality Trees

The last example of a root cause analysis tool is the current reality tree. As was the case with the interrelationship digraph, this tool is covered briefly. This methodology is similar to that of the some of the other RCA tools examined thus far.

The most obvious difference is that it is oriented vertically, so that the root problem starts at the bottom and the root causes show up in the top of the chart. Figure 7.8 shows a current reality tree for the employee who slipped in the supply room and broke his leg.

Timelines

No matter which root cause analysis methodology is used, the investigation should start by putting together a timeline of events. It can be a simple text list of dates, times, and events, or a graphical depiction of events, but get the information together as quickly as possible when gathering the baseline information for the investigation.

It is common to have people make comments such as, "This happened about the same time as this problem, so they are probably related," only to find out that the problem occurred before the event, so they could not possibly be related as cause and effect. In the previous example, you might find out that the floors had indeed been mopped and had been slippery,

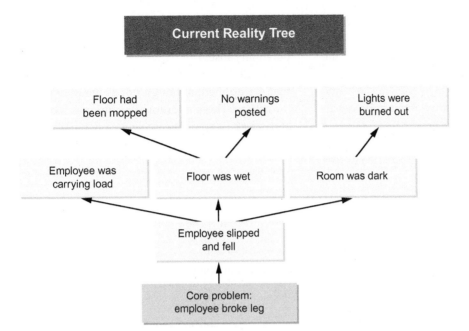

Figure 7.8 Example of a current reality tree.

but that this had occurred several hours before the slip, so the floors would have had plenty of time to dry. The timeline in Figure 7.9 shows the events related to the employee who slipped in the supply room and broke his leg. There is no substitute for good data, and a timeline helps provide good data.

Figure 7.9 Example of a timeline.

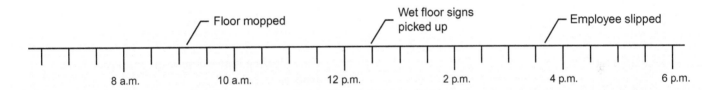

Summary

The purpose of root cause analysis is to determine the true underlying cause of a problem so that its recurrence can be prevented. Recognize that most of the time there is not a single root cause but a list of contributing root causes that may include actions, conditions, and circumstances. Many root cause analysis teams will require the direct participation of process technicians in order to understand the problem.

Before you would be asked to lead or facilitate a root cause analysis team, your employer would provide more detailed training on its particular methodology, but this chapter may be all the information you receive in root cause analysis before you are asked to participate in such an effort. The root cause analysis methodologies most commonly employed in the process industries include the Kepner-Tregoe approach, the Five Whys approach, and the Apollo RCA approach. Each methodology brings a slightly different perspective to the analysis.

The Kepner-Tregoe approach uses the "what it is" versus the "what it is not" approach to understanding the problem. The analysis of these contrasting bits of information helps to narrow down the list of potential root causes to only those that can satisfy all of the documented criteria.

The Five Whys approach suggests that for each cause that can be identified, you treat that cause as a problem and ask yourself why this new problem occurred. You must repeat this approach at least five times to get down to the true root cause.

The Apollo root cause analysis approach builds on the Five Whys methodology by evaluating each potential cause for both the actions and the conditions that resulted in the cause. This recognition that there are usually combinations of actions and conditions that contribute to a problem is realistic and explains to some extent the popularity of this approach in the process industries today.

Each of these approaches requires a solid, detailed definition of the problem and the validation of the suspected causes before you can declare the true root cause. A root cause that has not been validated is not a root cause at all but, rather, a potential cause.

Finally, we reviewed a few other root cause analysis tools that are applied, although not as commonly, in the process industries. These additional techniques included the cause-and-effect diagram, the interrelationship digraph, the current reality tree, and timelines as tools for analyzing data.

Checking Your Knowledge

1. Define the following key term:
   a. Root cause analysis (RCA)

2. Which of the following root cause analysis techniques is best described as using the contrast between what is versus what is not in its methodology?
   a. Five Whys
   b. Apollo
   c. Kepner-Tregoe
   d. Current reality tree

3. Which of the following root cause analysis techniques is best described as a chain of multiple questions digging deeper and deeper into the cause of the problem?
   a. Five Whys
   b. Apollo
   c. Kepner-Tregoe
   d. Current reality tree

4. Which of the following root cause analysis techniques is best described as seeking out the combination of actions and conditions that converge to cause a problem?
   a. Five Whys
   b. Apollo
   c. Kepner-Tregoe
   d. Current reality tree

5. Which of these is common to all root cause analysis techniques?
   a. Need data to validate the true root causes
   b. Need a solid definition of the problem
   c. Need to apply a team approach to the problem
   d. All of the above

6. The following is an example of which root cause analysis technique?
   a. Kepner-Tregoe
   b. Cause-and-effect diagram
   c. Current reality tree
   d. Interrelationship digraph

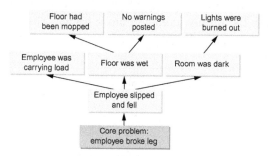

7. The following is an example of which root cause analysis technique?
   a. Kepner-Tregoe
   b. Cause-and-effect diagram
   c. Current reality tree
   d. Interrelationship digraph

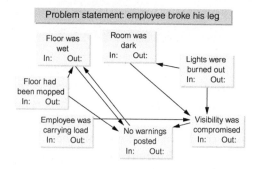

8. The following is an example of which root cause analysis technique?

    a. Kepner-Tregoe

    b. Cause-and-effect diagram

    c. Current reality tree

    d. Interrelationship digraph

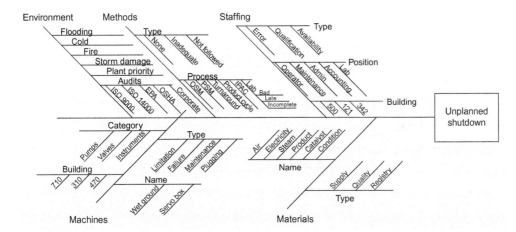

9. (*True or False*) Process technicians will rarely be called upon to participate in root cause analyses.

10. (*True or False*) Determining the true root cause is essential if you are to prevent the recurrence of the problem.

11. (*True or False*) Root cause analysis tools are relatively new and as yet untested in the process industries.

12. (*True or False*) A timeline of events is often useful in reconstructing a problem so that you know when a problem occurred and when it did not.

13. (*True or False*) Validation of the suspected cause(s) with data is essential to determining the true root cause.

14. (*True or False*) There may be more than one root cause to any given problem.

15. (*True or False*) Graphical tools such as the cause-and-effect diagram can be used to guide a root cause analysis team.

NOTE: Answers to Checking Your Knowledge questions are in the Appendix.

Student Activities

1. Several weeks ago, a mandatory evacuation was required for your area. Unfortunately, traffic conditions were horrible, and it took you almost 8 hours to evacuate. Working in teams of three to five students, conduct an Apollo root cause analysis for this problem. The expected outcome from this effort includes the following:

    a. A problem definition statement

    b. An Apollo style cause-and-effect diagram

    c. Identification of possible cause(s)

    d. An explanation of the types of data needed to validate these possible causes

    e. A short list of true root causes

2. The process industries continue to see more new builds taking place in foreign countries than in the United States.

Working in teams of three to five students, conduct a Kepner-Tregoe root cause analysis on why this is the case. The expected outcome from this effort includes the following:

    a. A problem definition statement

    b. An analysis of "is" versus "is not"

    c. Hypotheses of possible causes

    d. An evaluation of your hypotheses against the "is" and "is not" criteria

    e. A listing of probable root causes (those that meet the "is"/"is not" criteria)

    f. An explanation of the types of data needed to validate these probable causes

    g. A short list of true root causes

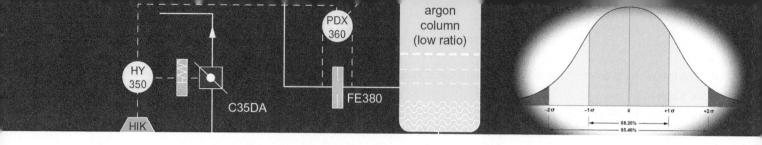

Chapter 8

Continuous Improvement— Six Sigma

"Excellence is not a destination; it is a continuous journey that never ends."

~ BRIAN TRACY

 Objectives

Upon completion of this chapter you will be able to:

8.1 List and explain the steps of the plan-do-check-act (PDCA) cycle. (NAPTA Quality, Variance and Operating Consistency 6*) p. 103

8.2 Describe the concepts and development of Six Sigma and the participant structure. (NAPTA Quality, Overview 7) p. 105

8.3 Describe the stages of a Six Sigma project and the key improvement tools used at each stage. (NAPTA Quality, Overview 7) p. 108

8.4 Explain how Six Sigma tools could be applied to new designs as well as existing processes. (NAPTA Quality, Overview 7) p. 115

Key Terms

Black belt—team leader on a Six Sigma team who has training in advanced quality improvement tools, plus training in project management, **p. 107**

Champion—in Six Sigma terms, a manager responsible for the overall effectiveness of the Six Sigma implementation at a business or functional level, **p. 107**

*North American Process Technology Alliance (NAPTA) developed curriculum to ensure that Process Technology courses will produce knowledgeable graduates to become entry-level employees in process technology. Objectives from that curriculum are named here in abbreviated form. For example, "(NAPTA Quality, Variance and Operating 6)" means that this chapter's objective 1 relates to objective 6 of the NAPTA curriculum on variance and consistency.

Design for Six Sigma (DFSS)—the application of appropriate Six Sigma tools to new process designs rather than to the improvement of existing processes, **p. 115**

Green belt—team member on a Six Sigma team that has training in the basic quality improvement tools, **p. 107**

Master black belt—Six Sigma expert who has training in many advanced quality improvement tools and is responsible for coaching black belts and providing training to green belts and black belts, **p. 107**

Plan-do-check-act (PDCA) cycle—a cycle for continuous improvement that implements the following steps: planning, doing, checking, and acting. Some companies use plan-do-study-act or a version called standardize-do-check-act (SCDA), **p. 103**

8.1 Introduction

The plan-do-check-act (PDCA) cycle is used with many basic quality tools. This chapter gives a brief overview of that continuous improvement cycle.

Six Sigma sounds statistical, and the lowercase Greek letter sigma (σ) is used in statistics to represent the standard deviation. You will learn more about standard deviation in Chapter 13, *Variance and Operating Consistency*. This chapter, however, is not about statistics. It is about an improvement strategy that uses statistics to accomplish improvement, but it is much more than a collection of statistical tools.

In this chapter, you will learn about the various stages of the Six Sigma improvement strategy and what tools are employed during each stage. Six Sigma can be applied to manufacturing processes as well as functional work processes. It can be applied to existing processes as well as new processes. It can even be applied in the design phase before a product or process is launched.

Plan-Do-Check-Act (PDCA) Cycle

Walter Shewhart proposed that quality improvement is accomplished through a **plan-do-check-act (PDCA) cycle** (Figure 8.1). As its name would suggest, the PDCA cycle consists of a set of repeating steps.

Plan-do-check-act (PDCA) cycle   a cycle for continuous improvement that implements the following steps: planning, doing, checking, and acting. Some companies use plan-do-study-act or a version called standardize-do-check-act (SCDA).

Figure 8.1 The plan-do-check-act (PDCA) cycle.

CREDIT: ananaline/Shutterstock.

First, you have to *plan* your attack. What is the current situation? What data exist to support the issue? What opportunity is there for change? What does the process look like?

After studying the situation and analyzing the data, you can *do* something about it. In the doing phase, you implement solutions to accomplish improvements to the system.

In the *check* phase, you re-analyze the data to validate that the improvement worked. Did the improvement eliminate or only reduce the problem? How much of an improvement did it make? Is further improvement needed or has there been enough?

From these questions, you decide how to *act*. If you have completely eliminated the problem or reduced it to an acceptable level, then you can start again in the plan phase to work on the next issue.

If the first pass through the cycle failed to address the issue or made a reduction but did not reduce the problem to an acceptable level, then you will need to revisit the plan phase to decide what the next approach will be. Either way, review of the results of the initial improvements should always culminate in starting over again.

Continuous improvement means always looking for ways to make processes and products better. You can never stop moving forward or you will most certainly be left behind. It is safe to say that every company or operating unit has some area of opportunity for improvement.

The PDCA cycle is not the only strategy for accomplishing quality improvement. You will notice many similarities between the PDCA cycle and Six Sigma. Think of the PDCA cycle not as a tool to employ but as a guideline to follow while on your continuous improvement journey.

In your role as a process technician, you may be asked to participate in an improvement project for which the issue has already been identified. For example, you may serve on a team to reduce or eliminate unplanned shutdowns of the process. In that case, the job of the improvement team is to do the following:

Plan—collect the data around the unplanned shutdowns, analyze the data to determine the most common cause of shutdowns, brainstorm possible solutions, and select the most effective one.

Do—implement the selected solution, either all at once or beginning with a pilot or test case to make sure the solution really works. Implementation should include updating of the documented procedures and drawings to ensure the problem remains fixed.

Check—once the solution has been implemented, you will validate the results. Did you get all or most of the expected improvement? Were there other results obtained as a result of your implementation that were not expected? Often, your processes are made of many interconnected parts, and changing something at one point will cause a change that you were not expecting somewhere else.

Checking is an especially critical role for process technicians. As the employees out in the field day and night, they are the most likely to notice a small change in the process that may not be readily apparent through the normal control systems readouts. Many times, there are process technicians who really know their equipment who provide early warnings to problems by noticing small, abnormal changes. You have to trust the instrumentation for your process, but you should also learn to "feel" the process because it can sometimes tell you when things are happening that instrumentation does not measure. It can even tell you that the instrumentation itself needs attention.

Act—how you react to the results depends on what the results are. If the problem is fixed, then you can move on to the next problem. If the problem is not fixed to the satisfaction of the business, then you need to go back and try again. Do not be discouraged if the first pass at any given problem does not eliminate it. Often an issue has several contributing factors, and your initial solution may have addressed only one of them. In that case, your action is to revisit the first phase and analyze the data again. This time, because you have already made some improvement, you have more and better data to analyze.

Sometimes you learn that what you thought would solve the problem did not solve it. Even this scenario can be classified as a success if you take the time to incorporate the learnings into your policies, procedures, and drawings.

Thomas Edison was fond of saying that there were no failed experiments. With each experiment that did not achieve the desired result, he eliminated one of the possible options and narrowed down the field, getting one step closer to success (Figure 8.2).

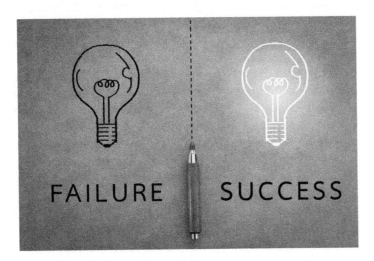

Figure 8.2 Success is often achieved after many failed attempts.

CREDIT: WHYFRAME/Shutterstock.

The PDCA cycle is never complete. PDCA is a strategy developed by quality experts Deming and Shewhart. Its premise is that quality is not a destination—it is a journey. It is what keeps a product alive and competitive in the marketplace.

8.2 Brief History of Six Sigma

In Chapter 1, *Introduction to Process Quality*, you were introduced to the term *total quality management* (TQM), which is the consolidation of many different components of quality into a single system. Six Sigma is an example of total quality management (Figure 8.3). It combines many elements of quality into a single management system.

Figure 8.3 The key elements of Six Sigma.

CREDIT: Trueffelpix/Shutterstock.

Motorola gets the credit for having "invented" Six Sigma in the mid-1980s. General Electric, Honeywell, Allied Signal, and Texas Instruments were some of the first companies to leverage Motorola's success. When Motorola won the Malcolm Baldrige National Quality award in 1988 and again in 2002 (see Chapter 4, *Quality Management System–International Standard [ISO]*), it attributed its success to its implementation of Six Sigma throughout the organization at all levels and in all functions.

All of these early adopters of Six Sigma were electronics manufacturers. These companies make thousands and thousands of discrete parts. In contrast, the process industries produce batches of process material. Perhaps because of this difference, it took the process industry a little longer to identify its need for tools like Six Sigma. Chemical giants Dow and DuPont began using the Six Sigma model in the late 1990s.

Again, thinking back to Chapter 1, *Introduction to Process Quality*, recall that the leaders in the quality movement had somewhat different approaches. Dr. Deming advocated statistical expertise at the master's level. Dr. Juran promoted project-by-project improvements using a breakthrough strategy. Phillip Crosby taught an attitude of zero defects and measurement of the cost of quality. Tom Peters spoke about the need for high-level quality champions to drive improvement down through the organization.

This book proposes that, instead of picking which of quality expert is "the best," it is more useful to adopt the best of each of expert's practices. This is exactly what Six Sigma has done as an improvement strategy.

Six Sigma drives improvement from the top by starting with management as the champion for the effort. It uses the language of management to quantify the value of the improvement. Six Sigma improvements are made by applying the appropriate statistical and graphical improvement tools in a team-based approach. Many of the tools used in the Six Sigma process are discussed in this book.

Six Sigma Defect Definition

Since it began, Six Sigma has grown and evolved from a metric to a methodology to a management system. In Chapter 16, *Process Capability*, you will learn how process variation relates to product specifications, using the process capability index (Cpk). You will also learn that ± 2 sigma is equal to 95 percent of the total variation of a normal distribution. If 95 percent of your process is inside these limits, that leaves 5 percent outside the limits. If your specifications are set at these limits, then your Cpk would be 0.66.

Table 8.1 shows that a 2σ process has a Cpk of 0.66 and a defect rate of 5 percent. Customers typically want a process that is less variable than that, often desiring a 4σ process, which equates to a Cpk of 1.33 and a defect rate of 63 parts per million (ppm). From Table 8.1 you can then see that a 6σ process has even higher standards. A 6σ (i.e., Six Sigma) process would have a Cpk of 2.0 and a defect rate of only *2 parts per billion (ppb)*.

Table 8.1 Six Sigma Defect Rate

Cpk	Sigma	Percent Passing	Defect Rate
0.66	2σ	95%	5%
1.00	3σ	99.73%	2,700 ppm
1.33	4σ	99.9937%	63 ppm
1.50	4.5σ	99.99966%	3.4 ppm
1.67	5σ	99.9999427%	0.5 ppm
2.00	6σ	99.9999998%	2 ppb

If you read about Six Sigma on the internet, you may see information about the process yielding a defect rate of 3.4 ppm and wonder why this differs from the calculations just mentioned here. The reason is that these calculations assume that the process is perfectly stable; thus, the mean of the process never changes.

The Six Sigma world has taken into account that the process *mean* (average) drifts around a little bit. It allows for a process drift of 1.5σ. This may sound a little confusing at first, but the gist of it is that a 6σ process is one in which a 1.5σ drift in the mean is allowed and the

IN A NUTSHELL

Six Sigma

A formal, disciplined, top-driven approach to continuous improvement using the best practices of the historically recognized experts in the field of quality.

process still has a capability of 4.5σ. This results in a Cpk of 1.5 or only slightly better than the targets historically employed in the process industry.

When a company makes hundreds of thousands or even millions of individual parts, even a relatively small defect rate can add up to a large problem. Let us say a company makes wood screws. It makes 50 million wood screws per year. A 4σ process (Cpk = 1.33) would mean a defect rate of 63 ppm or a total of more than 3,000 wood screws that are not sellable every year. On the other hand, because the company makes, let us say, 350 batches of product per year, a 4σ process means it would take 45 years to total one complete failing batch.

Let us put this in terms that are less industrial. A person plays 100 rounds of golf every year and averages two putts for each of the 18 holes in each round. That is 3,600 opportunities to putt per year, on average. A 2σ process is a 5-percent defect rate. Missing 5 percent of the putts means the golfer misses 180 putts per year or about two putts per round. Improving the process to 3σ means missing only 10 putts per year. If the golfer could achieve the process industry target of 4σ (Cpk = 1.33), that person would miss a putt only once every 4.5 years.

The purpose of this discussion is to put defect rates into perspective and to illustrate why the electronics industry sets goals so much higher than the process industry does. A defect rate of 5 percent is not too bad if you make only 10 batches per year, but it is completely unacceptable if you make 10 batches per hour. Most golfers would be happy to establish a 3σ process (Cpk = 1.0) and miss only one putt in every 10 rounds of golf they play.

Six Sigma Participant Structure

The Six Sigma structure is patterned after the martial arts. Process technicians are often trained as **green belts** in the process industries. Team leaders have further training and are referred to as **black belts** (Figure 8.4). A black belt might manage 2 to 10 green belts on a given project team.

Green belt team member on a Six Sigma team that has training in the basic quality improvement tools.

Black belt team leader on a Six Sigma team who has training in advanced quality improvement tools, plus training in project management.

Figure 8.4 The Six Sigma program is structured like the levels of karate, so a Six Sigma master is a black belt.

CREDIT: wavebreakmedia/Shutterstock.

Master black belts may oversee any number of black belts, providing training and assistance as needed. In Six Sigma, as in the martial arts, progression through the ranks is accomplished by training and demonstration of the required competencies. Some companies may include additional rankings, but those mentioned above are universal in Six Sigma applications.

In addition to these rankings, Six Sigma practitioners recognize the role of management in the overall improvement effort. Sometimes managers are certified as green belts or black belts and serve on improvement teams as leaders or participants. In their decision-making roles, managers may also serve as champions. As stated earlier in this chapter, a top-down commitment is required for success.

To ensure this high-level commitment, many companies identify business or functional **champions**. The role of the champion is to identify and select high-priority projects, ensure

Master black belt Six Sigma expert who has training in many advanced quality improvement tools and is responsible for coaching black belts and providing training to green belts and black belts.

Champion in Six Sigma terms, a manager responsible for the overall effectiveness of the Six Sigma implementation at a business or functional level.

that training is available to the team members, and remove roadblocks that might prevent the team from being successful.

8.3 Stages of Six Sigma

Six Sigma is a process for continuous improvement. As with most processes, it is made up of multiple steps or stages. Each of these stages has a purpose to fulfill for which there is a variety of tools available. Some tools can be applied in more than one stage; others are specific in their application. Before you study a more detailed description of the stages of Six Sigma, you need to understand a few basic concepts:

- Top-down deployment is an absolute must. To be successful, Six Sigma must be driven from the top of the organization. Upper management must champion the cause—not just support it. They must be fully committed. Commitment is often gauged by the level of active participation. At Motorola, vice presidents were actively participating in team activities.

- Process technicians probably will not be the ones to decide whether Six Sigma will be implemented. However, if a company has implemented Six Sigma, it is a great opportunity to participate and help make the program successful. Learning about the tools and methodologies in Six Sigma is a real advantage because the tools are time-tested winners.

- Key measures are best quantified in dollars. As discussed in Chapter 2, *Total Quality Management and Economics*, money is the reason companies are in business. In most organizations, there are more opportunities to improve the process than there are resources and staff to do so. Defining the opportunity in terms of dollars helps management ensure that it is working on the strategies that will help the most.

- The voice of the customer is key to success in the planning (define) stage. If the improvement is important to the customer, then the improvement is important. If the customer will not recognize the improvement, then it is not clear why the company would be working on the issue.

- $Y = f(x)$. Do not let the mathematical formula scare you. This is read as Y is a function of x (as seen later in Chapter 10, *Group Problem Solving—Designed Experiments*). Y is the output, and x is the collection of inputs). This statement can be restated as, "Our output is a function of our collective inputs." One premise of Six Sigma is to focus on the inputs in order to drive the desired output.

With these basic concepts in mind, we move on to the stages of Six Sigma: define, measure, analyze, improve, and control (DMAIC).

Define or Charter Stage

Some publications include *define* or *charter* as the first stage of Six Sigma; some do not. Even if it is not included in the official process, somebody has to define what the project is before any work is done. Management has to make a decision about where to deploy its resources and charter a team with responsibility to accomplish improvement. Although a process technician might not be involved in the define/charter step, companies with superior vision are recognizing the importance of their process technicians (operators). Safe, reliable operation of billion dollar process units is routinely in the hands of these employees every hour of the day and every day of the year. For TQM to permeate an organization successfully, it has to be top driven as mentioned, but it must also be pushed by line employees. Process technicians might be involved in the define/charter step and could be enlisted to identify processes where TQM would be profitable. More and more, today's degreed process technicians can expect a level of across-the-board employee involvement.

Project Charter

What is wrong?
(Define the Defect)

What needs to be done about it?
(Goals and Objectives)

Where does it need to be done?
(Project Scope)

Who are the right people to get it done?
(Team Membership)

How long should it take?
(Timeline)

How will we know when it is finished?
(Deliverables)

The main output from the define stage is a project charter. The charter answers some basic questions about the improvement effort. This is where management defines the defect, establishes the strategic importance, assigns resources to work on the project, defines the scope of work, establishes a desired timeline, and documents the required deliverables. Successful accomplishment of the team's goals will require planning up front to ensure that the right people with the right skills and training are there, and they are given the right amount of time to solve the problem.

Consider a project team that has been handed a charter with the following defect definition: Energy costs in the dibutyl ether unit are too high. The goal for the team is to deliver improvements to the process that will reduce energy costs by 50 percent. That sounds acceptable, but consider this: The consumption of utilities has remained fairly flat over time, but the price of the same utilities has risen by over 300 percent in the last 4 years.

This project is assigned to a production department team including salaried and hourly people. The plant engineers and process technicians may be able to make significant gains in energy efficiency and lessening consumption rates, but they probably do not have any control over utility costs. In this case, management is absolutely right in seeing a need to reduce energy costs, as these costs relate directly to the bottom line. The problem is that the goals and objectives of the team are outside management's span of control.

In this example, the project team could implement some elegant solutions such as reusing heat from one step of the process to preheat another step, shortening cycle times, or installing solar-panel-powered controller equipment. This team might be able to reduce energy consumption by 50 percent but still not meet its goals due to factors outside its control.

One of the first things a project leader should do when assigned a new project is review the charter with management to ensure the team membership is capable of delivering results that meet the goals and objectives of the project within the timeline provided. Although arguing with management is not recommended, tactfully challenging the premises of the charter to ensure success is definitely in the best interests of both management and the team.

Measure Stage

The main objectives of the measure stage are the following (Figure 8.5):

1. Finalize the project charter.
2. Document the current process.
3. Define the key input variables.
4. Establish the measurement system capability.
5. Establish baseline measurement of the defect.

Figure 8.5 Measure stage.

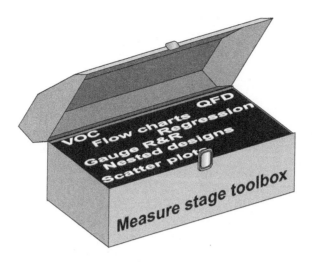

Once a good project charter has been documented, a reasonable way to kick off the project team is to review the charter with the team to gain a consensus understanding of what is going to be done, who is going to do it, and how they are going to do it. Chapter 6, *Team Skills*, can be reviewed for its coverage of team dynamics and types of teams.

The next activity is to document the current process. Flowcharts (see Chapter 11, *Other Basic Quality Tools*) are a useful mechanism for documenting a process graphically. If everybody on the team has a different view of what the process looks like, then the team will never come to an agreement on how to fix the problem assigned to it. A flowchart can establish a common level of understanding among the participants. In some cases, the act of mapping out the current process shows the team not only what the problem is but also how to fix it. Use whatever flowcharting, procedure review, or process mapping you desire, but never skip this step. One excellent technique for process mapping is the Rummler-Brache approach. This approach would be excellent reading for anyone interested in going a step or two beyond the basics.

Management has given the Y in the equation $Y = f(x)$. It has specified which output is in need of improvement. The job of the team is to define which process inputs are the ones that are generating that output. The team has to define the xs. The answer may be fairly obvious and require no more effort than reviewing the process flowchart and writing down the answer. In the energy consumption example just talked about, it should be straightforward to define where the company uses steam, where it uses electricity, and so on. But consider a project in which the Y is customer satisfaction.

Knowing exactly why customers are not satisfied is not typically quite as obvious. Sometimes even customers themselves cannot really put a finger on the problem; the customer just does not like doing business with you. Review Chapter 3, *Customer Service*. The team might be able to make some guesses about why customers are unhappy, but they can only work on the right xs if they collect data to measure customer satisfaction.

The tools most often used to collect customer data are called voice of customer (VOC), quality function deployment (QFD), and the quality design plan portion of quality reliability planning (QRP), which was discussed in Chapter 5, *Quality Management–Quality Reliability Planning*.

If a defect is related to product quality, the team must measure the amount of poor-quality material being produced and determine what process parameters are leading to poor quality. This will require use of statistical tools such as scatter plots, regression analysis, and designed experiments, which are discussed in other chapters.

Once the output (Y) and the inputs (xs) are known, you need to understand how to measure them (Figure 8.6). Everything varies. If you do not think it varies, it is just because you cannot measure the variation. If you cannot obtain a "good" measurement of inputs and outputs, then there is no way to improve upon current performance.

Figure 8.6 You must study and understand your variables.

CREDIT: ilkercelik/Shutterstock.

Another tool that accomplishes this goal is called the gage reproducibility and repeatability (gage R&R) study. If the team finds itself in the awkward position of being charged to make an improvement in an area that cannot be adequately measured, it may have to spend some time in this stage improving the measurement system or even putting a new measurement system in place. Improvement cannot be made if it cannot be measured.

The final piece of work for the measure stage is establishment of baseline performance using the accepted measurement practices. Refer back to one of the statements made earlier about project charters: "Reduce energy consumption by 50 percent." The team has to establish some basis against which to measure this improvement. Is that a 50-percent reduction from last month's performance or from last year's average performance? Once the baseline performance data are established, you are ready to move on to the next stage of the Six Sigma improvement strategy.

Occasionally, a project ends at the measurement stage. It is possible for a team to establish the measurement system capability and collect baseline data, only to discover that the magnitude of the problem was not what had been expected. In that case, it is appropriate to thank the team for its efforts and disband it to reassign the resources to more pressing matters. This is not a failure. This is good management, using the data to drive decision making.

Analyze Stage

The main objective of the analysis stage is simple: to determine the root cause(s) of the problem (Figure 8.7). This can be accomplished by doing the following:

1. Analyzing the data
2. Brainstorming potential causes
3. Validating causes
4. Selecting root cause(s)

Analyzing data that have been collected is the obvious starting point. This analysis may take the form of control charting, histograms, analysis of variation (ANOVA), or a myriad of other statistical techniques.

The team will typically brainstorm a list of potential root causes and then use the appropriate statistical analyses to validate or invalidate each potential cause. Many of the tools mentioned in the measure stage are employed in this stage as well. For example, designed

Figure 8.7 Analyze stage.

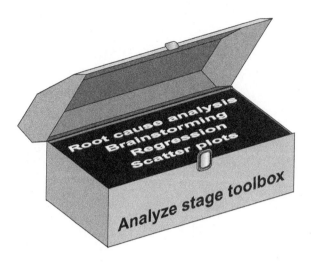

experiments and regression analysis are commonly used to establish a relationship between the inputs and the output. In addition, the root cause analysis (RCA) tools that were covered extensively in Chapter 7, *Continuous Improvements–Root Cause Analysis (RCA) and Corrective Action/Preventive Action (CPA)*, are commonly employed:

- Cause-and-effect matrix
- Five Whys
- Kepner-Tregoe root cause analysis
- Apollo root cause analysis.

If all goes well, the list of process inputs correlates to the output that is the focus of the project. This correlation validates which process parameters are root causes of the output and which ones are not. Oftentimes, the analysis of the data is so closely tied to the collection of the data that the analyze stage is completed quickly after the measure stage.

Once the causes of the problem have been identified, you can progress to fixing the problem. There is no requirement to use each and every tool mentioned above. What is required is to document the valid root cause(s). The team should feel free to employ whatever tools that help it get there. Many teams make the mistake of jumping directly into fixing the things that "everybody knows" to be the problem, without going through the disciplined root cause analysis. This method typically leads to fixing a known problem but not fixing the problem assigned to the team.

Improve Stage

The objectives of the improve stage (Figure 8.8) are all focused on eliminating or reducing the root cause(s) determined in the preceding stage by doing the following:

1. Brainstorming potential solutions
2. Evaluating potential solutions
3. Selecting optimum solution(s)
4. Validating optimum solution(s)

Sometimes the team identifies a single root cause that can be eliminated, driving the defect down to zero. More often the team identifies multiple contributing causes, each of which contributes to the defect. Sometimes these can be eliminated, but sometimes they can only be reduced.

Imagine, for example, that there is raw material contamination in the plant. If you have the ability to test the raw material before it is used in your production process, you may be able to determine a root cause and to eliminate the contamination. If the contaminant is unknown or you do not have the technology to detect the contamination, you may not be

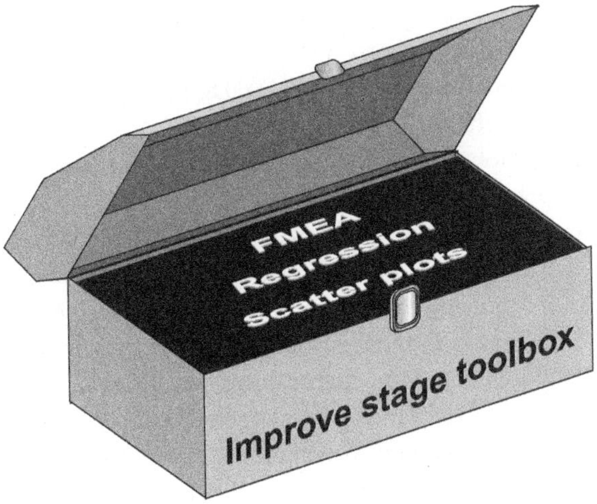

Figure 8.8 Improve stage.

able to eliminate the root cause. Improvements at the company that supplies the raw material may help reduce the problem, but this is outside your company's scope of control and may not be completely dependable.

The first step is generating ideas about how to fix the problem. Once the root cause or causes are known, it is usually straightforward to think of ways to fix the problem. The goal of Six Sigma at this stage is to challenge the team to think of possible solutions that are less obvious. Just as there may be more than one cause, there may be more than one solution. Sometimes the best solution is made up of several different ideas that build on each other.

With a wide range of options identified, the team then needs to evaluate the potential solutions. A matrix can be created to rate each solution against selection criteria that it agrees to in advance. In discussions with the process owner, the team determines that the selection criteria include the following issues:

- Cost to implement
- Cost to manage after implementation
- Time to implement
- Impact on the project objectives (low, mid, high)
- Complexity
- Probability of success

The team can set limits around what constitutes low cost versus high cost, define short time to implement versus long time to implement, and so on. With these criteria defined, each potential solution can be evaluated to see which option looks most promising and efficient. If management places a higher importance on cost to implement than it does on time to implement, then the team can use a weighted average to give more weight to those issues related to cost. The end result is a short list of solutions that are expected to achieve the most benefit for the least amount of money in the shortest amount of time. With management's help, the team selects the solutions to implement.

Most of the time, a full-scale implementation of a huge project will not gain upper management approval without being tested on a smaller scale. This is called a *pilot program*. The company may be willing to spend thousands of dollars to make sure the project will work, in order to reduce the risk of spending millions of dollars to find out that it has a critical flaw in the logic.

In one case, a major chemical company was led to believe that a new style of blender was required to achieve the quality targets for a certain product line. This new style of blender was estimated to cost between $15 million and $20 million. Based on the research conducted by the team, this particular style of blender had been used extensively in the food processing industry in making bulk chocolate, but it had never been used before in the chemical industry.

The team was able to identify a small-scale blender that would mimic the commercial scale, but its purchase and installation cost was over $1 million. In this case, management chose to spend approximately $10,000 to set up a laboratory scale test using rented equipment and hired technicians to validate the concept. Based on the positive results, it agreed to spend the $1 million to implement a small-scale pilot facility capable of making large enough quantities of product to send to its customers for testing. When the customer feedback was positive, management then authorized the implementation of the $20 million commercial-scale solution.

Sometimes the right answer is to rent the solution. Sometimes the right answer is to buy a smaller-scale model first. Sometimes the team may be able to run the existing equipment using different procedures or different conditions. In each case, the team must not only evaluate the beneficial impact of the solution but also identify any possible negative trade-offs. It would not make sense to implement a new procedure that achieves significant reductions in energy usage but that makes more off-spec product than before or that has increased maintenance issues due to operating equipment above or below design specifications. In other words, you have to be sure you are not trading one problem for another one that is potentially worse.

If the team is able to use the existing equipment, then the validation may be as simple as running a few experiments to prove the concept. One excellent tool for evaluating potential failure modes is the failure mode and effects analysis (FMEA), which was covered in some depth in Chapter 5, *Quality Management–Quality Reliability Planning*.

The result of the improve stage is a list of validated solutions that will fix the problem assigned to the team.

Control Stage

The final stage of the Six Sigma process is the control stage (Figure 8.9), the objective of which is to ensure that the desired improvements are maintained. Many have heard seasoned employees in industry describe a problem that had to be fixed again and again because the organization implemented what is called a temporary fix instead of a permanent fix.

Figure 8.9 Control stage.

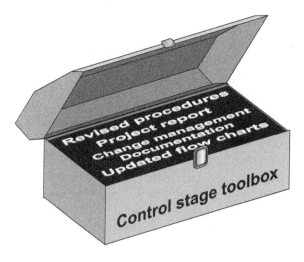

Consider a project team that determined that running the reactor at 100 degrees Celsius (212 degrees Fahrenheit) would provide a 10 percent quality improvement over the current operating condition of 90 degrees Celsius (194 degrees Fahrenheit). How should the team implement this change? Administrative controls work well, but there is no way to guarantee that they will stay in place.

- Instructing all four shifts of process technicians to run at 100 degrees Celsius (212 degrees Fahrenheit) instead of 90 degrees Celsius (194 degrees Fahrenheit) is good.

- Updating the procedure manual to state that the desired operating temperature is 100 degrees Celsius (212 degrees Fahrenheit) is better.

- Changing the process/alarm set point in the process computer so that the change happens automatically is better still.

- Changing the setpoint in the process computer, updating the procedure manual with the new requirement and an explanation of why the change was made, and training all four shifts of process technicians on the nature of the change is the best solution yet.

One thing that gets overlooked far too often is the "why" of the change. If the reason for change is not made clear, then the organization may soon forget and revert to the old conditions when a new team made up of people who either did not participate or cannot remember the reason for the change is chartered to work on a different problem.

The tools employed at this stage of the project are things such as revised procedures, updated flowcharts, a formal project report, and some type of change management documentation. If a company has a control plan for each product, then that quality control plan (QCP) should be updated and a new FMEA constructed as well. (The QCP and FMEA were both discussed in more detail in Chapter 5, *Quality Management – Quality Reliability Planning*.)

The last action that the improvement team needs to take is to transfer ownership of the project back to the process owner. The team can draft the updated procedures, revise the flowcharts, and fill out all kinds of project documentation. However, if the owner of the process does not take ownership in the change, the benefits will not be maintained. The transfer activity is often accomplished by holding a formal meeting at which the process owner has an opportunity to review and accept the work of the improvement team.

8.4 Design for Six Sigma

The Six Sigma approach outlined in this chapter is the standard version. It is used extensively throughout the process industries to accomplish the improvement of existing processes. Another type of Six Sigma that can be found in industry is called **Design for Six Sigma (DFSS)**, which applies Six Sigma to new processes. Unlike the usual Six Sigma—which is universally accepted as the define, measure, analyze, improve, and control (DMAIC) methodology—DFSS comes in many different varieties. Rather than explore the many types of DFSS in existence, this chapter will describe DFSS in general and fairly brief terms.

DFSS is used in the design phase of a new product or process to ensure that the process does what it is supposed to do from the beginning, negating the need for improvement after startup. The idea is to design something right the first time, so it does not have to be redesigned later. Many of the tools employed are the same tools as are used in the DMAIC variety of Six Sigma. Flowcharts, brainstorming, cause-and-effect matrices, and others are all employed. Because everyone is starting from scratch, there is a strong need for tools such as voice of the customer (VOC) and FMEA in Design for Six Sigma. The statistical tools, such as designed experiments, require the use of data that often do not exist in a new application. Instead, models generated from existing processes are sometimes used to predict the operation of a new process, so even these tools find use in DFSS.

It may seem that design engineers, not process technicians, would make up a DFSS team. Wrong. Process technicians and engineers working in processes similar to the one being designed are in the best position to review the proposed design and point out potential trouble spots. A good design team will take advantage of a process technician's expertise if given the chance.

Design for Six Sigma (DFSS) the application of appropriate Six Sigma tools to new process designs rather than to the improvement of existing processes.

Summary

This chapter discussed the plan-do-check-act (PDCA) cycle, which is a is designed for continuous improvement by performing a series of steps: planning, doing, checking, and acting.

The main topic of this chapter is Six Sigma. The chapter compares the tenets of Six Sigma to earlier quality efforts such as total quality management. Notice that Six Sigma represents the best practices of each of the historically recognized quality experts, pulling these pieces together into a single improvement strategy.

Six Sigma, as an improvement strategy, allows for a 1.5 sigma shifting of the mean. Because of this allowance, a process that is said to be a Six Sigma process is one that will produce defects at a rate of 3.4 per million opportunities.

Six Sigma practitioners are named using a martial arts–type convention with green belts being team members, black belts being team leaders, and master black belts being the implementation experts and trainers. In addition, many companies identify the role of champions (managers who are responsible for the overall effectiveness of the Six Sigma program).

Six Sigma is made up of multiple stages defined as the define, measure, analyze, improve, and control (DMAIC) methodology. Within each of these stages, the discussion identified the objectives and outlined the types of improvement tools used as follows:

Stage	Objectives	Tools
Define	Project charter	Charter template
	• Defect definition	
	• Strategic importance	
	• Goals and objectives	
	• Project scope	
	• Team membership	
	• Timeline	
	• Deliverables	
Measure	• Finalize the project charter	Gage R&R
	• Document the current process	Nested designs
		Quality function deployment
	• Define the key input variables	Voice of the customer
	• Establish the measurement system capability	Regression analysis
		Scatter plots
	• Establish baseline measurement of the defect	Flowcharts

Stage	Objectives	Tools
Analyze	Determine root cause by:	Root cause analysis
	• Analyzing the data	Brainstorming
	• Brainstorming potential causes	Designed experiments
		Scatter plots
	• Validating causes	Regression analysis
	• Selecting root cause(s)	
Improve	Eliminate root cause by:	Designed experiments
	• Brainstorming potential solutions	Scatter plots
		Regression analysis
	• Evaluating potential solutions	Failure mode and effects
		Analysis
	• Selecting optimum solution(s)	
	• Validating optimum solution(s)	
Control	Ensure that the gains are maintained	Quality control plan
		Failure mode and effects
		Analysis
		Flowcharts
		Revised operating procedures
		Change management
		Documentation
		Project report

Finally, the traditional application of Six Sigma to improve existing processes was contrasted with a relatively newer application of some of the same tools to the design of new products and processes. The application of Six Sigma tools to new designs is called Design for Six Sigma.

As a process technician, you can expect to participate in the Six Sigma process, most probably as a green belt. Some of the tools you have learned or will learn in this textbook. Other tools will be taught to you once you join the workforce.

Checking Your Knowledge

1. Define the following key terms:
   a. Black belt
   b. Champion
   c. Design for Six Sigma (DFSS)
   d. Green belt
   e. Master black belt
   f. Plan-Do-Check-Act (PDCA) cycle

2. (*True or False*) The plan-do-check-act (PDCA) cycle ends after the process technician has acted on the problem.

3. The PDCA cycle is a strategy developed by which quality experts? (Select all that apply.)
   a. Shewhart
   b. Deming
   c. Juran
   d. Baldridge

4. Team members on a Six Sigma project are called:
   a. green belts.
   b. black belts.
   c. master black belts.
   d. champions.

5. Team leaders on a Six Sigma project are called:
   a. green belts.
   b. black belts.
   c. master black belts.
   d. champions.

6. Six Sigma experts responsible for providing training in quality improvement tools are called:
   a. green belts.
   b. black belts.
   c. master black belts.
   d. champions.

7. Six Sigma can be best described as:
   a. a metric used by managers to judge conformance to requirements.
   b. a formal, disciplined, top-driven approach to continuous improvement.
   c. a project team.
   d. a Greek letter that means standard deviation.

8. A Six Sigma process yields a defect rate of:
   a. 2 defects per billion opportunities.
   b. 10 putts per year.
   c. 64 defects per million opportunities.
   d. 3.4 defects per million opportunities.

9. (*True or False*) Six Sigma is not well integrated into the process industries.

10. (*True or False*) Six Sigma uses the best practices of the historically recognized quality gurus.

11. Which of these is NOT a stage of Six Sigma?
    a. Analyze
    b. Measure
    c. Design
    d. Control

12. Determining the root cause is the main objective of which stage?
    a. Measure
    b. Analyze
    c. Improve
    d. Control

13. Eliminating the root cause is the main objective of which stage?
    a. Define
    b. Analyze
    c. Improve
    d. Control

14. Ensuring that the organization maintains the gains made is the main objective of which stage?
    a. Measure
    b. Analyze
    c. Improve
    d. Control

15. (*True or False*) Six Sigma improvement tools can be applied only to existing processes.

16. (*True or False*) The intent of design for Six Sigma (DFSS) is that it will eliminate the need for improvement after startup of a new facility.

NOTE: Answers to Checking Your Knowledge questions are in the Appendix.

Student Activities

1. Suppose that your class is struggling with the concepts of quality. Draft a project charter for an improvement team.

Project Charter
Title:
Defect Definition:
Strategic Importance:
Goals and Objectives:
Project Scope:
Team Membership:
Timeline:
Deliverables:

2. Go to the official Six Sigma website (in your internet browser, type: six sigma). At the top of the page, click the Explore dropdown, and click Recent Articles. Choose one article about Six Sigma; read it and take notes on it. Give a short presentation on the article to the class, explaining why you chose this article and what it described.

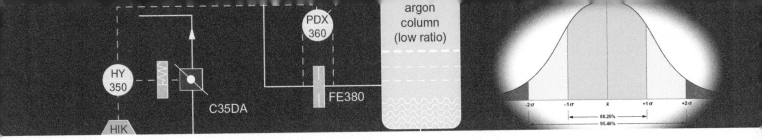

Chapter 9

Continuous Improvement—Lean

"Lean manufacturing is at the core of our strategy to support growth. You have to take waste out of everything. We only want to do things that our customers are willing to pay for."

~ DAVID BUCK

Objectives

Upon completion of this chapter you will be able to:

9.1 Define Lean manufacturing as used by process industries. (NAPTA Quality, Continuous Improvement 3*) p. 120

9.2 Discuss specific Lean tools used in process industries:

Value stream maps

Kaizen

5 Ss

Just-in-time production. (NAPTA Quality, Continuous Improvement 1, 3) p. 123

Key Terms

5 Ss—a floor-level improvement methodology taken from five Japanese words: Seiri, Seiton, Seiso, Seiketsu, and Shitsuke, **p. 128**

Continuous improvement—an ongoing effort to improve products, services, or processes using critical thinking skills, **p. 123**

Just-in-time (JIT)—process by which the company creates a product for the customer when the customer needs it, **p. 129**

*North American Process Technology Alliance (NAPTA) developed curriculum to ensure that Process Technology courses will produce knowledgeable graduates to become entry-level employees in process technology. Objectives from that curriculum are named here in abbreviated form. For example, "(NAPTA Quality, Continuous Improvement 3)" means that this chapter's objective 1 relates to objective 3 of the NAPTA curriculum on continuous improvement.

9.1 Introduction

The improvement strategy called *Lean* is heavily influenced by Japanese improvement efforts. Because of the Japanese influence, Lean is full of terms that make it more difficult to understand than other systems. In order to familiarize you with Lean as it is implemented in industry, this chapter will introduce you to many Japanese terms. However, you will concentrate on the concepts of Lean and will not have to learn Japanese to understand it or to succeed in this class.

Lean can be implemented as a stand-alone improvement strategy or as an add-on to an existing Six Sigma effort. Some of the tools in the Lean toolbox include value stream maps, Kaizen, 5 Ss, and just-in-time production. This chapter will introduce these tools and the Lean manufacturing practices used in the process industries.

What Is Lean?

Think of your process as a lean, mean quality machine. Everything works smoothly. There is no wasted effort and no fat (Figure 9.1). Efficiency is at an all-time high. That is exactly the concept of Lean—no waste. Waste costs money. If you can eliminate the waste, then you get to keep the money. Think of waste as an expense that reduces overall profit.

Figure 9.1 The concept of Lean. Bars of trash depict waste. Width of the trash bars is reduced as Lean waste reduction strategies are implemented.

CREDIT: Courtesy of Willie L. Myles. Background photo Flying object/Shutterstock.

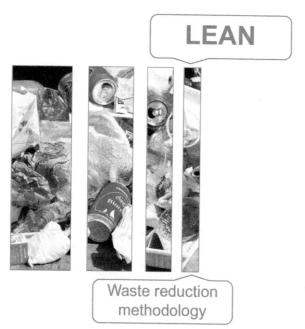

Many times, managers are busy managing and do not know how to transform their operation into a waste-free, lean operation. They must rely on the "workers on the floor"—process technicians—to accomplish this. These employees may not have an opportunity to implement Lean strategies if management does not support it, but Lean really is a floor level, or control room level, effort.

To explain the point about management support a little further, Lean is a philosophy or a culture more than it is a collection of tools. The tools are intended to be implemented at all levels of the organization, but the tools must be implemented in a supportive environment that starts at the top. If you do not have a supportive environment, then you will not have a Lean culture to maximize the success of implementing its tools.

Recall from Chapter 1, *Introduction to Process Quality*, that many of the tools in the field of quality were developed by Americans, but the success of the tools was realized as they were implemented by the Japanese during their rebuilding effort following World War II. Lean is an exception to this rule. Lean actually started out in the Toyota Motor Corporation over 60 years ago in Japan. Toyota called its quality improvement process the Toyota Production System, or TPS. The terms *TPS* and *Lean* are synonymous in the quality industry today.

In the 1980s, Dr. James Womack published a book about the TPS and called its quality operation *Lean*. The Toyota Production System was designed so that Toyota used fewer suppliers, had quicker turnaround times, produced product with fewer defects, stored less inventory, and in general had fewer wasted efforts than were common in traditional assembly line operations.

In the Toyota Production System, and in Lean today, waste is divided into three categories: Muda, Mura, and Muri.

Muda

Muda, which is a traditional term in Japan to describe waste, is further broken down into these seven subcategories:

Muda Japanese term for waste.

- Transport—in a production facility, if operation A is finished on one side of the plant and the product has to be moved to the other side of the plant to continue operation B, then the time it takes to move the product from A to B is wasted transport time. No value is added to the product during this time.

- Waiting—when the product arrives at operation B, it has to sit and wait for someone to pick it up and do something with it. There is no value added to the product while it is sitting and waiting.

- Overproduction—if the customer needs 1,000 pounds of product and the company makes 1,200 pounds to ensure that nothing goes wrong, then it has incurred waste. Employees are working to make product not designated for immediate sale and for which there is no immediate need. These employees could have done something that added value instead.

- Defects—obviously, making a defective product is wasteful. A great portion of this book deals with tools and techniques employed to analyze, detect, and reduce defects that waste time, materials, and resources. Good product will then have to be made to replace the defective product, consuming even more time, materials, and resources. Employees will have to expend additional effort dealing with the defective product. Even throwing away a defective product costs time and money.

- Inventory—inventory is necessary because it is what a company sells. Having more inventory than is needed is wasteful. Companies pay taxes on the inventory they carry. They pay for storage facilities to store the inventory and pay for people to manage the storage facilities. In short, companies pay money when they make product that sits on the shelf. In the Toyota Production System, companies make only what they need and make it when they need it.

- Motion—excess motion is another form of waste. For example, let us say a job on an assembly line is to pick up headlights from a box on the floor and install them on a chassis at chest level, and a worker installs 250 headlights on a good day. This means that part of the work is comparable to touching your toes 250 times a day. This excess motion is not only a waste of time, but it could also lead to repetitive motion injury (Figure 9.2).

Figure 9.2 Repetitive motion injury.

CREDIT: **A.** Marcin Balcerzak/Shutterstock. **B.** Ralf Geithe/Shutterstock.

A.

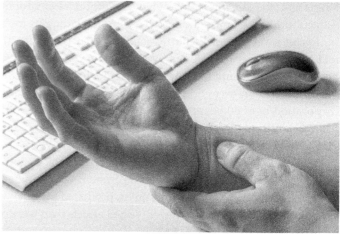

B.

- Excess processing—in general, any process step that does not add value to the product is, by definition, waste. Some argue that excess processing is obvious and should be eliminated from the list of subcategories. However, a surprising number of operations contain excess processing. What is done is what has always been done, even if nobody can say why it is done that way.

There are some instances in which waste is necessary, such as inventory (Figure 9.3). Oftentimes in the process industry, specialty products can take days or even weeks or months to produce. Companies cannot wait for customers to demand the product and then tell them to wait a few weeks while it is produced. Retail businesses must have inventory sitting on the shelf; otherwise, the customer would have no reason to shop in their store.

Figure 9.3 Inventory is often a necessary type of waste.

CREDIT: TaraPatta/Shutterstock.

Another example of necessary waste is in the excess processing area. Sometimes there are process steps that do not add value to the product, but they help the company with regulatory compliance or address some environmental or safety issue. For example, toxic off-gases are treated so they will be safe to vent, even if that part of the process does not add value to the customer. Even though all transport, motion, inventory, and so on are waste, companies accept that some waste is necessary. The job of Lean is to determine which waste is unnecessary and to eliminate it.

Mura

Mura may be described as unevenness. Americans often call this *variation* or *lack of uniformity*. Variation is the enemy of quality. Variation is a classification of waste. When it occurs, employees must expend time and energy to measure the product to determine how much variation exists and whether the product can still satisfy the customer. If the product is made exactly the same every time, then these wasteful steps can be eliminated. Variation is the focus of Chapter 13, *Variance and Operating Consistency*, Chapter 14, *Variables Control Charts and Interpretation* and Chapter 15, *Attributes Control Charts and Interpretation* in this textbook.

Some process technicians work in laboratories or have small field laboratories associated with their processes. By the Lean definition, all laboratory effort is a form of waste, made necessary by variation in processes and products. Eliminating variation means that wasteful laboratory work can also be eliminated.

Mura Japanese term for unevenness.

Muri

Muri is the Japanese term to describe unreasonable burdening of employees. This is the hardest term for Americans to grasp. It has to do with asking too much of employees or processes. Employees who want to make extra money often consider working overtime a good thing, but working too much overtime can tire workers out, cause family problems, and increase the probability of safety incidents happening on the job.

Muri is the waste caused by making excessive demands on employees. Consider the individual who had to pull a 16-hour shift during the startup of a new process. The task was completed, but the work done during the last 4 to 5 hours of that shift was not that person's best.

Muri also relates to ergonomic issues. Companies pay more attention to ergonomics now than in years past, making it more comfortable for employees to get their jobs done while reducing safety and health issues caused by repetitive motion, strain, and so on.

Lean tools are actually much easier to work with than some other more sophisticated tools. Many of the tools in this book are designed to achieve a great step change in the performance of the operation. In contrast, Lean focuses more on incremental improvements at the shop-floor level. The simplicity of the Lean tools and the ability to implement them at the worker level have contributed to its growth in popularity in the process industries.

Muri Japanese term for overburden.

9.2 Key Lean Tools

How do companies go about making their organizations Lean? Dozens of tools are available in the Lean toolbox (Figure 9.4), and they all focus on **continuous improvement** and

Continuous improvement an ongoing effort to improve products, services, or processes using critical thinking skills.

Figure 9.4 Lean toolbox.

elimination of waste. This text presents four of the most commonly applied tools or concepts associated with Lean:

Value stream maps

Kaizen

5 Ss

Just-in-time production.

Keep in mind that Lean is a culture of optimization that can be widely applied, so the tools you apply will depend on the operations in which you work. If you are in the supply or logistics area of the plant, then managing inventory to keep it at an absolute minimum will be the focus. If you are in the manufacturing area of the plant, then minimizing defects may be the focus. Laboratory workers might focus on transport and motion waste in order to perform inspection with as little unnecessary waste as possible.

Value Stream Maps

Value stream map (VSM) a high-level flowchart of a process. Included within each step of the process is the amount of time and the value it adds from the customer's perspective.

The foundation of Lean tools is the value stream map. A **value stream map (VSM)** is a high-level flowchart that puts a value on each step that a process adds from the customer's perspective. Figure 9.5 is a simple rendering of a VSM. In this figure, the process starts in the upper right-hand corner. The customer wants product, so an order is placed. Each part of the process is broken down into hours. Storage time, though necessary, is not considered value-added time.

Figure 9.5 Example of a value stream map showing the time involved in each step of the process.

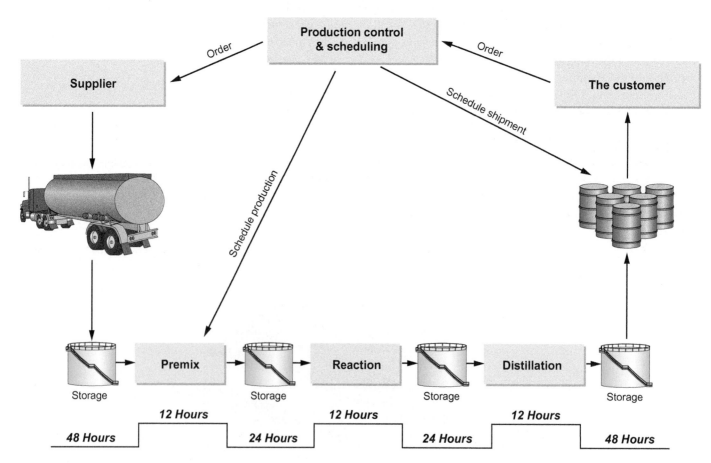

An actual VSM may include dozens of process steps. Each step may identify not only the amount of time, but also other measurements, such as personnel requirements, utility requirements, or inventory loading. The process starts in the upper right-hand corner, where the customer wants product and places an order.

The production control and scheduling center must order raw materials from the necessary suppliers and schedule the production facility to make the product. If the company has the product on hand, then it might ship the product from the warehouse to the customer while making product to replenish the warehouse.

The process is broken up into several different sections. Each one pulls material from storage, conducts some processing step, and then puts that material into inventory to have available for the next step in the process. When the product is completed, it is packaged and put into inventory to await the next customer order.

Say, for example, that it takes 48 hours from the time you contact the supplier until the time you have raw materials on hand. It takes you 12 hours to prepare your premix, which is a blend of the right amounts of materials for the product. This premix is put into storage for an average of 24 hours so the reaction group can use it when needed. Reacting the premix with the catalysts and other bulk ingredients is a 12-hour process.

The product that you make in the reaction step is called the *raw product*. The product is made but it is still too dirty to sell. The raw product goes into storage for the next processing step. The average storage time is 24 hours. The raw product is then fed through a distillation column where the impurities are removed and the product is purified for sale. This step takes approximately 12 hours.

The final product is then pumped to a large storage tank, where it awaits the drumming operation. It takes approximately 48 hours to get the product drummed, labeled, analyzed, and delivered to the warehouse, where it can finally be sold to customers.

From the time your customer wants product to the time the product is delivered to the warehouse for shipment is 180 hours—7½ days. Because your customers demand delivery within 48 hours of placing an order, you must maintain inventory on hand in order to meet their demands. In this example, only 36 hours of the total 180 hours was value added time. Eighty percent of the time was time spent waiting on equipment to be ready or waiting for the product from the previous step of the process.

Let us say the warehousing costs for this operation are a million dollars per year. That is a million dollars that could be saved by eliminating the need to carry inventory. To implement the Lean system, a company could start by scheduling production so that each step of the process is ready to take material from the previous step as soon as the previous step is finished. The supplier may be asked to maintain a storage tank on the premises and manage the raw material inventory for the company. Then the company can pull inventory whenever it is needed and keep records to balance the books at the end of the month. An automated drumming machine could be installed at the end of the process to fill drums and palletize them for shipment directly to the customer.

In a perfect world, a company would not need a warehouse for this example. Because the world is not perfect, some inventory may need to be kept on hand. However, that inventory might be able to be reduced from an 8-day inventory to a 2-day inventory. This would be a significant saving.

With the Lean system, the idea is to match the time it takes to make product to the consumption rate of customers so that the process is perfectly balanced and timed (Figure 9.6). This rate of production to meet customers' need is called **takt time**, a German term that comes from musical tempo or beat ("keeping time"). Think of takt time like this: If every part of the process is rowing to the same beat, then it is an efficient, lean, well-run process.

Assume that a company has 12 hours of production available per day. The customer demand is 360 drums per day. The takt time would be 1/30 hour, or 2 minutes per drum. The facility would need to produce a pallet of 4 drums every 8 minutes in order to keep up with this customer's demand.

Takt time a German term that refers to keeping a beat. In quality discussions, it is the available production time divided by the customer demand.

Here is an everyday example using the concept of takt time to pace a process. Imagine a cookout at a church that plans to serve hamburgers at noon to over 200 people. In order to get that much food ready by noon, the head cook plans to start grilling hamburgers at 9:30 A.M. The grill holds 20 patties, and it takes about 15 minutes to cook one batch. In order to have 200 patties ready by noon, there need to be about 80 patties cooked per hour, and it would take 2½ hours to prepare all the patties.

Of course, cooking so early would mean the hamburgers would be cold by the time they were served. To prevent this, the next step is to apply the takt time concept. It takes approximately 1 minute for each person to go through the line to get meat, bread, fixings, chips, drink, and dessert, so the cook needs to take a patty off the grill every 1 minute. It takes about 15 minutes to grill a burger, so the first patty needs to start being cooked 15 minutes before noon. The next patty would go on the grill at 14 minutes before noon, and the next one at 13 minutes before noon.

Since the grill can hold about 20 hamburger patties at a time, there is plenty of production capacity to have at least 15 patties in various stages at any point in time. A patty goes on every minute and one is taken off every minute. Using the takt concept, every person in the line of 200 people can get a hot juicy hamburger without the cook doing anything other than pacing the process to match the customer need.

An important thing to consider at every single step of the process is whether a step is value added. To be value added, a process step must do the following:

- Add value from the perspective of the customer (the customer must be willing to pay for it).

- Transform the product in some way.

- Be defect free (doing something over again is not value added).

All three conditions must be met; otherwise, the step is considered some type of waste.

As with most of the improvement strategies and tools that are covered in this book, the work process is to apply the tools to the current state, analyze the information, make improvements based on what you have learned, then reapply the tool to assess the results. In the case of VSM, it means making a *current state map*, perhaps drawing a future state map to guide the improvement process, and then drawing the resulting improved process map, which becomes the new current state map.

When making value stream maps, be sure to talk to the people who work in all parts of the process, and be sure to walk through the process yourself. It is amazing how many times problems can be solved when individuals observe the process to see what is happening.

The Japanese word for "where the action is" is *gemba*. The Japanese phrase for the concept of going to where the action is to see for yourself is *genchi genbutsu* or "go and see."

A value stream map can show where improvement opportunities lie, but it does not say how to improve the process. That is why this tool and the concept of Lean are classified as an improvement strategy. To employ this strategy requires use of capability studies, control charts, designed experiments, and Pareto charts. The tools covered in Chapter 7, *Continuous Improvements–Root Cause Analysis (RCA) and Corrective Action/Preventive Action (CPA)* and Chapter 8, *Continuous Improvement–Six Sigma* fit together within the Lean improvement strategy to accomplish the improvements that everyone seeks.

Kaizen

Kaizen is another term from the Japanese that continues to be used in discussions about Quality. It describes small, steady, positive changes. Many of the tools discussed in this book are focused on making a step change difference in the performance of a process. In contrast, Kaizen focuses on incremental changes that can be made quickly and cheaply on a day-by-day basis.

Kaizen Japanese term for continuous improvement or incremental improvement.

Did You Know?

The principle of Kaizen is Japanese:

No matter what you do or where you go, leave the place in better shape than you found it.

This idea is not uniquely Eastern. Campers and hikers have a similar motto: Leave the campsite better than you found it.

KAIZEN

CREDIT: NirdalArt/Shutterstock.

As an employee, every day you should look for opportunities to clean up around you and make small, incremental improvements to your processes. Do not go home until you have made your work environment a better place to work. This is the concept of Kaizen.

One of the basic tenets of Kaizen is that you should employ the plan-do-check-act (PDCA) cycle, which was discussed in Chapter 8, *Continuous Improvement–Six Sigma*. Kaizen projects can be carried out by individuals, by work group teams, or by management teams. Some improvements need to be made at a high level in the company that only management can accomplish. Other improvements can be accomplished by individuals or by small, self-directed work group teams.

Much attention is given in the Lean world to the concept of Kaizen, but it all boils down to this: While working on the big breakthrough improvement projects, do not forget to work on making everything you touch just a little bit better every day. It is a culture, a way of life. It is about respecting the people around you, the environment, and the workplace.

In his book *Kaizen*, Masaaki Imai emphasizes the differences between Western philosophy and Eastern philosophy. Eastern Lean philosophy emphasizes small, constant improvements, but Western managers like big improvements. A manager will likely not get a bonus for reorganizing the control room to make running the process easier and more efficient, yet doing so improves the process. It is difficult to transplant culture from East to West, but the concept must be accepted in order to apply the philosophy of continuous improvement.

Five Ss

5 Ss a floor-level improvement methodology taken from five Japanese words: Seiri, Seiton, Seiso, Seiketsu, and Shitsuke.

The concept of **5 Ss** is straightforward and can be used by anybody at any level in the organization. The 5 Ss is a methodology taken from five Japanese words. They are listed here, followed by an attempt to translate the Japanese 5 Ss into English:

1. Seiri (SAY-REE)—sorting, organization
2. Seiton (SAY-ton)—straightening up, neatness
3. Seiso (SAY-SO)—sweeping or scrubbing, cleaning
4. Seiketsu(SAY-ket-dzuh)—standardizing, standardization
5. Shitsuke (shit-tsuh-KEH)—self-control, discipline

STEP 1: SORT OR SEGREGATE   Get organized. Everything in your office, on your workbench, or in your closet can be segregated into categories:

- Items that you use all the time
- Items that you use occasionally
- Items that you never use
- Items that are not even usable.

In order to apply the 5 Ss, you have to sort through everything in your area and segregate it into these categories.

STEP 2: STRAIGHTEN UP   Put items used all the time close by, where they can be accessed easily. Put items that are used occasionally farther back. Put items that are hardly ever used all the way in the back. Get rid of items that are not used at all or are unusable.

A good example of seiton organization is a computer keyboard. The home position (the place where fingers are supposed to stay while typing) has the index fingers over the F and J. Many keyboards put tabs on the F and J keys so you can find them by touch without looking down. With hands in the home position, the letters that are used the most often in the English language are under or around your index, middle, and ring fingers. Letters that are used the least often are farther from the center. The keyboard is organized to put what you use most into the most accessible places.

Another example is in the organization of tools. Many mechanics and hobbyists organize their tools on pegboards to make them easy to find and easy to put away. There should always be a place for everything, and everything should be in its place (Figure 9.7).

Figure 9.7 Tools organized so each has a space.

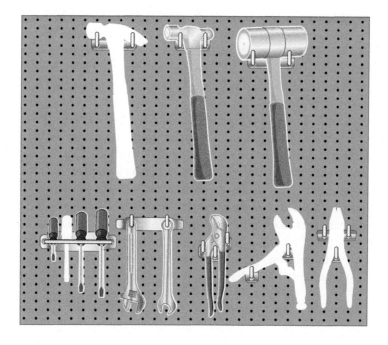

STEP 3: SWEEP OR SCRUB   Clean up the place. Most companies call this housekeeping. A clean place to work is a Lean place to work. One look at the pegboard in Figure 9.7 indicates that there are a several tools missing. In this step of the 5 Ss, you find them and put them where they belong. Otherwise, when you need them, you will not know where they are.

STEP 4: STANDARDIZE   This is a step that occurs over and over again in the field of quality. Chapter 4, *Quality Management Systems—International Standard (ISO)*, discussed standards in depth. Many process technicians will work in companies that operate on shifts, which means there will be three or four people working the same job in the same location using the same facilities during different hours of the day. If everyone has a different way of doing things, everyone will be wasting time and energy looking for procedures or looking for equipment or undoing something somebody else has done. If everyone does things in the same way, then all of the procedures, tools, and equipment will always be in the right place.

STEP 5: SELF-CONTROL   Discipline. The goal here is to maintain the gains. Having a fancy pegboard for tools does not do any good if you never put the tools back on it. If you have a team of four process technicians who rotate shifts and three of them are disciplined but one is not, then the whole organization will suffer. The person on the shift following the undisciplined technician will be frustrated that things are not how they should be. You would not want to work with an undisciplined technician, so plan not to be one.

Just-in-Time and Kanban

The traditional logistics process in America today is a push system. Teams of commercial and marketing experts forecast what they believe sales will be in the coming months, and production is scheduled to meet the forecasts. These forecasts are used to *push* the production schedule. To ensure the company can meet this projected demand, product is made in advance of orders, and inventory is stockpiled.

As you have already learned, inventory is waste, and the whole idea of a Lean organization is one without waste. In a Lean organization, you would wait for the customer actually to need the product and pull the product from you. You would make the product as requested, and there would be no stockpiled inventory. The product would be made **just-in-time (JIT)** to meet the customer's needs.

It is complex for processing industries to implement JIT production because of the lead time that is often needed to produce the products, but it can be done. For example, one major chemical manufacturer makes a specialty catalyst product used around the world. A typical batch of product takes 4 to 6 weeks to manufacture and another 2 weeks to blend, package, and test.

In order to manufacture this product, the company must order a custom-produced raw material from another state. This raw material takes 4 to 6 weeks to manufacture. To produce the raw material, the supplier must order base materials that can take 6 weeks to prepare. Add up all the lead time, and you will see that from the time the customer orders the product, it could be 4 to 5 months before it is ready to ship. Depending on where the customer is located, shipping time could add another month.

Amazingly, this particular organization does not carry any stock inventory. The customers and the producing plant have put in place a system of communications and reviews that ensures orders are placed with at least 6 to 8 months of lead time. This is just-in-time manufacturing. The customers pull the product out of the organization when they need it, with minimal wasted time and inventory.

Some writings about JIT manufacturing refer to the concept as **Kanban**, the Japanese word for card or sign. The technical application of Kanban to JIT manufacturing is as follows: When there are several steps in a process, the product must flow from step to step within the processes. Kanbans are the inventory management cards used to signal that more product is needed from the previous step in the process. Literally, they are the signal used to pull inventory forward into the next step of the process. As outlined above, JIT has been successfully implemented in the process industries. Kanban cards are mentioned here because you may be part of an organization that uses them.

Just-in-time (JIT)   process by which the company creates a product for the customer when the customer needs it.

Kanban   a signal card used to pull production.

More Lean Terminology

Here are a few more Japanese words used in Lean manufacturing that you should at least be able to recognize:

- *Andon*—a signal to alert people to a problem. A red light that flashes when someone pulls the fire alarm is an andon. The "check engine" light in a car is another example of an andon.

- *Jidoka*—making automated machines smart enough to stop a process when something is wrong. Many processes now have sensors that stop a process when it is not flowing properly, thus preventing it from making defects. In the real world, a breathalyzer can be a jidoka device. When installed in a car, it can read a person's blood alcohol content and prevent the car from starting if it is over the limit.

- *Poka-yoke*—a device or system that prevents defects by error proofing the process. For example, a diesel-dispensing nozzle cannot be used by mistake in a gas-powered car because it is bigger than the opening for an unleaded gasoline tank. The process has been error proofed.

- *Sensei*—master or teacher. Just as the Six Sigma process refers to practitioners as green belts and black belts, Lean leaders are called sensei.

Lean Six Sigma

Lean Six Sigma combines aspects of the Lean and Six Sigma improvement strategies (Figure 9.8). Although they are different, they are complementary. Lean Six Sigma's goals are to eliminate waste and inefficiency, to resolve problems, and to improve working conditions in order to provide a better response to customers' needs. It combines:

- Tools and techniques for identifying and solving problems
- Phases for organizing, finding root causes of problems, and implementing full solutions
- A mindset of reliance on data and process to achieve continuous improvement.

This team-oriented approach to improvement has shown some success in maximizing efficiency and improving business profitability.

Figure 9.8 Lean Six Sigma.

CREDIT: Courtesy of Willie L. Myles. Background photo: Flying object/Shutterstock.

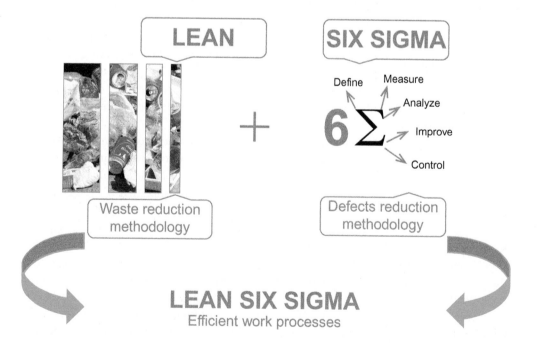

LEAN

SIX SIGMA

Define Measure Analyze Improve Control

6 Σ

+

Waste reduction methodology

Defects reduction methodology

LEAN SIX SIGMA

Efficient work processes

Summary

Having a Lean system means there is no fat or waste in an organization. Waste can fall into three larger categories:

- Wastefulness
- Variability
- Overburden.

Traditional waste can be any of seven different types: transport, waiting, overproduction, defects, inventory, motion, or excess processing.

The simplicity of the Lean tools and the ability to implement these tools at the worker level have contributed to the popularity of Lean in the process industries.

Key process improvement methodologies of Lean include value stream maps (VSMs), Kaizen, the 5 Ss, and just-in-time manufacturing.

The value stream map, or VSM, is a graphical technique used to analyze the steps of the process to show which steps add value and which steps do not. VSM can show where improvements are possible within the process and then help in applying the appropriate improvement tool to the process.

Kaizen is the Japanese word for continual improvement. Many of the statistical tools in this textbook drive toward a breakthrough or step change improvement. In contrast to them, Kaizen is the philosophy of making small, incremental improvements every day.

The 5 Ss are a simple way to accomplish the desired small, incremental improvements of the Kaizen philosophy. The English words for the 5 Ss are the following:

- Sorting
- Straightening up
- Sweeping or scrubbing
- Standardizing
- Self-control

Just-in-time manufacturing (JIT) is a mechanism used to reduce inventory, which is one of the seven types of waste. In JIT manufacturing, the customer pulls product when needed instead of using forecasts to push production schedules.

It is acceptable, when discussing the Lean culture, to talk about waste instead of *Muda*, to say "go to where the action is" instead of saying "*genchi genbutsu* to the *gemba*," and to talk about error proofing a process instead of *poka-yoke*. Lean is a philosophy of excellence. A person can learn and apply the philosophy without using all the Japanese terms.

Lean Six Sigma attempts to eliminate waste and inefficiency, to resolve problems, and to improve working conditions to better respond to customers' needs. It includes tools for identifying and solving problems, phases for organizing, identifying root causes, and applying solutions.

Checking Your Knowledge

1. Define the following key terms:
   a. 5 Ss
   b. Continuous improvement
   c. Just-in-time (JIT)
   d. Kaizen
   e. Kanban
   f. Muda
   g. Mura
   h. Muri
   i. Takt time
   j. Value stream map (VSM)

2. Lean manufacturing is best described as:
   a. a philosophy of working without waste.
   b. a plant with limited production hours.
   c. a plant with the bare minimum number of workers.
   d. a philosophy of working with the minimum amount of equipment.

3. What we call Lean today began in which automotive company?
   a. Ford
   b. General Motors
   c. Toyota
   d. Honda

4. Which of these is NOT a type of waste as defined in Lean terms?

   a. Muda

   b. Variation

   c. Overburden

   d. Overtime

5. Which of the following are types of Muda? (Select all that apply.)

   a. Transport

   b. Employee benefits

   c. Overproduction

   d. Motion

   e. Excess processing

6. A technique for analyzing a process graphically to look for opportunities to reduce wasted time is:

   a. just-in-time.

   b. value stream maps.

   c. Kaizen.

   d. flowcharting.

7. Takt time is:

   a. the rate of consumption to meet the customers' needs.

   b. the time required to rework a production error.

   c. the total process time divided by the number of customer orders per block of time.

   d. the ratio of time that does add value to time that does not add value.

8. To be considered a process step that adds value, which criteria must be met? (Select all that apply.)

   a. Customer willing to pay for the service

   b. Ensures compliance with government regulations

   c. Transforms the product

   d. Is free from defects

9. (*True or False*) All waste is unnecessary.

10. (*True or False*) Kaizen is a philosophy of small, incremental, continuous improvement.

11. (*True or False*) Just-in-time production means making the product when it is needed instead of making the product to put into inventory.

12. (*True or False*) Some of the Lean tools are so easy to use that they can be applied by anybody at any level in the organization.

NOTE: Answers to Checking Your Knowledge questions are in the Appendix.

Student Activities

1. Think back to the registration process for this class. Construct a value stream map of that process. What percentage of the total registration time was value added? Compare results in class.

2. Assess some area of your home, such as your workshop, your garage, or your closet. Can you see opportunities to apply the 5 Ss there? Describe what you would do in this area for each of the Ss. Share your findings with the class.

argon
column
(low ratio)

Chapter 10
Group Problem Solving– Designed Experiments

"Don't think of it as failure, think of it as designing experiments through which you're going to learn."

~ TIM BROWN

Objectives

Upon completion of this chapter you will be able to:

10.1 Explain the importance of modeling a process. (NAPTA Quality, Data Collection 1; Continuous Improvement 3*) p. 134

10.2 Determine the direction and strength of a correlation. (NAPTA Quality, Continuous Improvement 8) p. 134

10.3 Describe regression and the types of regression. (NAPTA Quality, Continuous Improvement 8) p. 139

10.4 Explain how designed experiments generate data that allow you to model a process. (NAPTA Quality, Group Problem Solving 1) p. 141

10.5 Explain the process technician's role in designed experiments. (NAPTA Quality, Group Problem Solving 1) p. 143

Key Terms

Correlation—the mutual relationship of two or more things; the degree to which two or more attributes tend to vary together, **p. 134**

Performance model—the mathematical model that predicts how your process will perform based on the input variables, **p. 134**

*North American Process Technology Alliance (NAPTA) developed curriculum to ensure that Process Technology courses will produce knowledgeable graduates to become entry-level employees in process technology. Objectives from that curriculum are named here in abbreviated form. For example, "(NAPTA Quality, Data Collection 1; Continuous Improvement 3)" means that this chapter's objective 1 relates to objective 1 of the NAPTA curriculum on data collection and objective 3 of the NAPTA curriculum on continuous improvement.

Regression—the statistical analysis that generates a mathematical representation of the correlation between two or more variables, **p. 139**

10.1 Introduction

At some point, process technicians encounter designed experiments in their work, so it is important to know how to contribute to their success. However, designed experiments are not the role of and are not overseen by process technicians. Discussion in this textbook is strictly about the concept of designed experiments; formulas are just provided to aid understanding.

In order to predict the future performance of a process, you need to have a model of it. This does not mean a plastic model that shows what the process looks like, although that type of model is also useful; it means a **performance model**. With a model in hand, it is possible to predict the outcome of changes that are made to the process control scheme. Without a model, you are basically flying blind and hoping everything will turn out all right.

How do we generate a model? We generate a model by using data from the process. Here is one of the cardinal rules of modeling: A model is good for predicting process performance only for the range of data that was used to calculate the model. In other words, if you ran your process pressure between 100 pounds per square inch (PSI) and 110 PSI and calculated a model for how much product your process generated, this model has been validated only between 100 and 110 PSI. You cannot trust that this model would be useful at 150 PSI.

It is quite common for processes to be held in tight control, so you may not have any data for 150 PSI. You might extrapolate what should happen as you go outside the range of 100 to 110 PSI, but you do not *know* what will happen. This poses a dilemma.

For your model to be useful, you need to collect data over as broad a range as is safely possible. However, running processes at widely varying conditions is completely the opposite of the consistent operation you try to achieve in order to produce a quality product. Designed experiments can be used to address this issue. Designed experiments are short, logically designed experimental runs whose purpose is to collect the greatest amount of useful data over the broadest range possible with the least amount of waste.

This chapter introduces the concept of designed experiments. It also introduces correlation and shows how to determine the direction and strength of correlations. It briefly discusses the use of regression analysis (the mechanism for turning data into models) and then covers designed experiments and the role of the process technician in making them work. An example of a small-scale Roman catapult will be used to demonstrate the concepts of designed experiments in a manner that can be applied in the classroom.

Performance model the mathematical model that predicts how your process will perform based on the input variables.

10.2 Correlation

Correlation is derived from "co" and "relate," meaning to have a connection or relationship together. A mathematical formula used with correlation follows.

$$Y = f(x)$$

This is read as "Y is a function of x." Y is process performance, the output of a process, and x is one of the inputs into the process. Let us say that a process is boiling water on the stove. Y is the temperature of the water in the pot, and x is the heat being applied to the pot. As you turn up the heat applied to your pot, you expect to get hot water. The temperature of the water is a function of the heat applied to the pot. Is the heat applied to the pot the only variable that affects the heat of the water? No. How much water is in the pot? How cold was the water when it was put into the pot? How long has heat been applied to the pot? Each of these variables could be measured, and the formula could be updated to be more precise. The new formula might take this form:

$$Y = f(v, w, x, z)$$

which you would read as "Y is a function of several variables." In this example, the heat of the water is a function of how much water there is, how cold the water was when heating started, how much heat has been applied, and how long the heat has been applied.

Correlation the mutual relationship of two or more things; the degree to which two or more attributes tend to vary together.

Have all of the possible inputs been named? There is almost always another variable that might help make the model better. One of the most respected industrial statisticians, George E. P. Box, is quoted as saying, "All models are wrong; some models are useful."

Correlation ≠ Causation

Consider these fictional graphs of the human population and the population of the common brown pelican in the United States from 1970 to 2000 shown in Figure 10.1.

Figure 10.1 Graphs of the U.S. population and the brown pelican population.

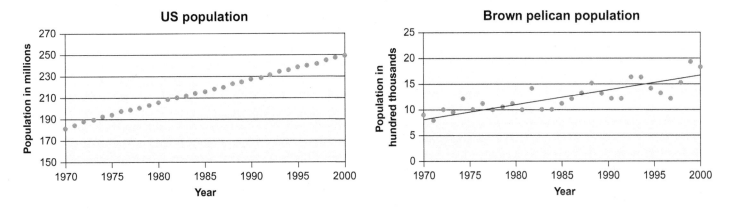

You can see from the graphs that each population appears to be increasing, so do we assume that they must be connected? In order to check for a correlation, the graph shown in Figure 10.2 was created.

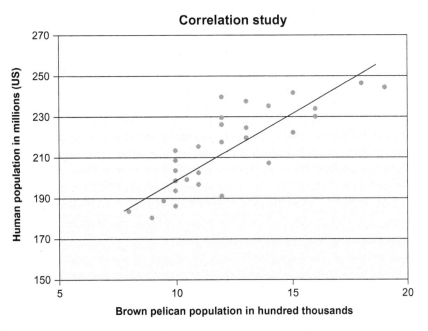

Figure 10.2 Correlation of the brown pelican population and the U.S. population.

It does appear that there is a relationship between the increase in human population and the increase in the numbers of brown pelicans. You can draw a line through the data that shows both populations increasing during this time span.

Next, we will compare the population of humans with the population of wood storks (Figure 10.3). (For anyone who is not familiar, storks are large, long-legged birds that sometimes roost near chimneys. This behavior gave rise to the folk tale that storks deliver babies by dropping them down chimneys into houses.)

Figure 10.3 Correlation of the wood stork population and the U.S. population.

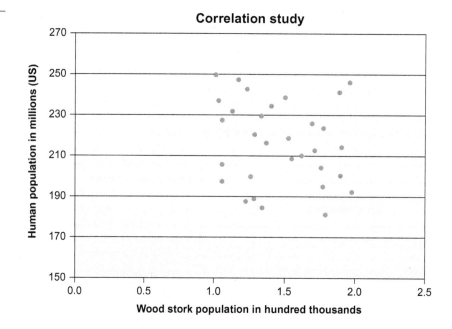

There is no identifiable pattern in this graph. In fact, it would be difficult to draw any conclusions at all from it. The graph suggests that there is no correlation between the human population and the stork population. Would these graphs lead you to conclude that humans are not delivered by storks but by pelicans?

Of course not. Correlation is not the same thing as causation. Our Figure 10.2 model shows a clear correlation between increasing human population and an increasing pelican population. That does not mean that pelicans deliver babies, but it does not mean that they do *not* deliver babies either. In this case, all that the data correlate, and all we can really conclude is that both populations were on the rise during this time frame.

As this scenario illustrates, blindly performing statistical calculations with a computer does not necessarily lead to more useful knowledge about a process. It is not performing the calculations that is important. Rather, it is what you do with the information you gain from the calculations that matters.

Sometimes engineers run experiments, gather data, and run statistical analyses to come up with impressive charts. To learn something from these analyses, everyone has to sit and stare at them for a while and ask themselves how meaningful they are. Does this correlation sound reasonable? Have any laws of physics been violated? Were other variables that were not included in the study changing at the same time? Process technicians play a key role in understanding what the results from these types of experiments mean. Oftentimes, they work night shifts or weekends, when there are fewer interruptions and it is easier to observe things that might be missed during the day.

Direction of Correlation

Now that you know what a correlation is, and what it is not, let us look at a couple of characteristics of correlations. Look at the direction of the correlation (Figure 10.4). If both properties go up together, a line starts at the lower left and goes up and to the right. These variables are said to be positively correlated.

Using the earlier example, if you want to boil water, you must increase its temperature. As the temperature of the water increases and gets closer to boiling point, the amount of time remaining to boil it is reduced. These two variables (temperature and time to reach boiling point) are negatively correlated. Negatively correlated variables will result in a chart that goes down and to the right.

Direction of correlation

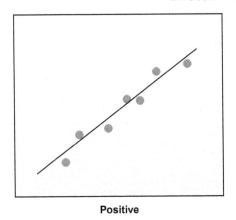
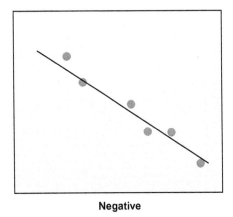

Positive   Negative

Figure 10.4 Positive and negative correlations.

Strength of Correlation

You can tell how strong a correlation is by graphing it (Figure 10.5). Let us look at the correlations between human population and the pelican and stork again.

Strength of correlation

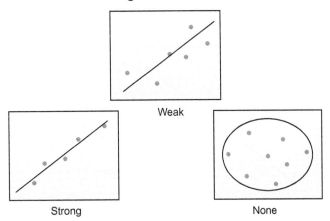

Weak

Strong   None

Figure 10.5 Strength of correlation.

The data from the pelican correlation make a straight line. The data from the stork correlation make a scatter plot. When you run the formal statistical analysis, you obtain a correlation coefficient that is a number between −1 and +1. A value of +1 means a perfect correlation in the positive direction. A value of −1 means a perfect correlation in the negative direction.

The closer to positive or negative 1 the correlation coefficient is, the stronger the correlation. The closer to zero the correlation coefficient is, the weaker the correlation. Generally, what you hope to find is a correlation coefficient in the 0.75 or higher range. This range securely establishes a link between the variables. A correlation coefficient less than 0.50 is not considered useful.

Dual Correlation

Sometimes there is a unique graph, such as Figure 10.6. In this case, you are attempting to correlate the product purity coming out of the reactor against temperature in the reactor. At first it appears there is just a poor correlation. However, while reviewing the results of the analysis, the process technicians working the board during the test run indicated they had run out of a certain raw material right in the middle of the run and had had to switch to the next batch in the warehouse.

Figure 10.6 Graph of
temperature versus purity.

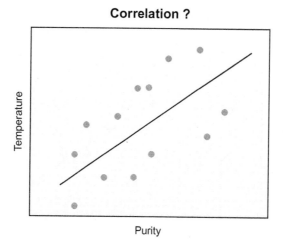

Given that clue, you color code the product made with raw material batch A differently from the product made with raw material batch B. With this additional information, it was clear that there was a strong positive correlation between the product purity and the reactor temperature (Figure 10.7), but the correlation was slightly different when using different batches of the same raw material.

Figure 10.7 Correlation determined from Figure 10.6 showing results from two different batches. Batch A data points are red. Batch B data points are blue.

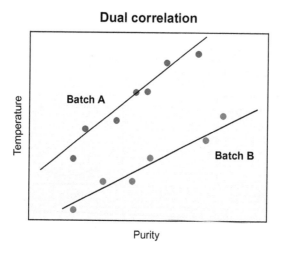

Had you not reviewed the test results with the test run crew, this phenomenon might not have been discovered. At the least, it would have a taken a lot more work to determine what had happened. In effect, you learned during this experiment that the product purity was a function of both raw material purity and reactor temperature. The next step was to examine the quality information on this raw material to see what was different between the two batches.

The results showed that one significantly different characteristic caused the difference in process performance. In this case, you were trying to develop a model for $Y = f(x)$, where x was the reactor temperature. Instead you found that you needed a model for $Y = f(x, z)$, where x was the reactor temperature and z was the raw material purity.

A simple correlation of product purity to reactor temperature was not possible. You ended up with two lines on the graph. Had three or four different batches of raw material been used, you would have ended up with three or four different lines on the graph. Clearly, there must be a better way.

Testing the parameters one by one could be endless, especially since you learned that the product purity versus reactor temperature line was not the same with different batch numbers of that particular raw material. What if the lines were different with other raw material variations as well? There could be an infinite number of raw material and process control combinations that would take your entire career to test.

Fortunately, it is not necessary to test one variable at a time. In fact, it is much better to learn how to test multiple variables at one time for the specific purpose of finding these interactions among variables.

10.3 Regression

The fitting of a line to the experimental data that have been collected is called a **regression** model. The quality of a regression is described by the correlation coefficient R-squared (R-sq or R^2). R-sq describes the fraction of the data explained by the regression. The higher the R-sq is, the better the regression explains the data.

Regression the statistical analysis that generates a mathematical representation of the correlation between two or more variables.

An R-sq of 1 would mean that 100 percent of the data is explained by the regression.

An R-sq of 0 would mean the 0 percent of the data is explained by the regression.

Therefore, the higher the R-sq, the better the model describes the process. Generally, an R-sq of 0.75 or higher is needed to have a model suitable for predicting process behavior.

Many types of regression analyses are used in different situations. The end result is that somebody plugs the data into a computer, and the computer cranks out an analysis that has to be interpreted. As has been discussed already, without the human interpretation of what the data mean and what the analysis appears to be indicating, you might draw all kinds of wrong conclusions (like that babies are delivered by pelicans instead of storks).

Simple regression is the correlation of a single x (input) to a single Y (output). It is the determination of the formula $Y = f(x)$.

Multiple regression is the correlation of many inputs to a single Y (output). It is the determination of the formula $Y = f(w, x, z \dots)$.

There are times when a simple regression is all that is needed. In the process industries, it is common to have complex operations with dozens of variables to study, so multiple regression analyses are most often employed.

As mentioned earlier in this chapter, the problem with using historical data to predict future performance is that your historical data may exist in too narrow a range. In order to overcome this issue, you can design experiments to collect a small amount of data over a wide range for each of the inputs and analyze those data to determine where to operate.

Did You Know?

German mathematician, Karl Friedrich Gauss (1777–1855) predicted the location of the dwarf planet Ceres, which had been discovered and then lost track of after only 41 days of observation. He took the observation data and invented the *least squares regression method* to predict the planet's new location at a given time. When astronomers looked where Gauss predicted, they found Ceres again.

CREDIT: Oleg Golovnev/Shutterstock.

Gauss worked from the assumption that Ceres's orbit was elliptical, with the Sun as a focus. He used his mathematical skill to calculate six theoretical quantities, which uniquely specified the size, shape, and orientation of the orbit in space. With these six theoretical quantities, the celestial position of Ceres could be calculated at any past or future time.

Let us look at some data from a chemical process. Process technicians filled in a log sheet every 2 hours for a couple of days (Table 10.1). Available data include the reactor temperature, the reactor pressure, and the yield, which is the measure of the process output.

Table 10.1 Log Sheet

Rx_Temperature	Rx_Pressure	Yield
50.4	6.8	87.5
52.6	9.3	93.2
51.3	8.9	90.9
53.0	7.3	91.7
52.7	6.3	90.4
52.9	5.2	89.5
52.8	6.5	90.7
52.8	7.4	91.5
51.1	6.7	88.4
50.8	8.6	89.8
51.5	9.3	91.6
51.2	5.7	87.6
50.3	8.2	88.6
51.4	7.3	89.4
50.1	9.8	89.9
51.3	8.7	90.6
52.6	6.6	90.5
50.2	9.5	89.8
52.0	5.5	88.5
50.3	9.6	90.1
50.8	6.4	87.7

The reactor temperature varied between 50 to 53 degrees Celsius (120 to 127.4 degrees Fahrenheit) during this time frame. The reactor pressure varied from around 5 to nearly 10 PSI. Under these conditions, the process produced between 87 and 92 pounds of product per hour.

Plotting the yield versus the temperature to see whether there was a correlation (Figure 10.8) did not result in much evidence of correlation. The yield versus the pressure was plotted as well (Figure 10.9). It appears that the same yield resulted no matter what pressure was run.

It is common knowledge that the temperature and pressure in the reactor are the control handles for the process, so where is the correlation? First, correlation is not evident because you are looking at one variable at a time, and these variables are related. Second, as noted in the preceding section, these input data are collected over a small range, the normal operating range.

The conclusions drawn from analysis of the data are only as good as the data being analyzed. This leads us to the topic of designed experiments.

Figure 10.8 Temperature and yield correlation.

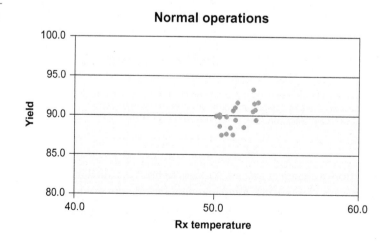

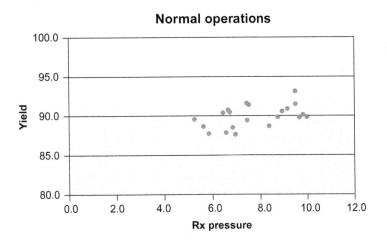

Figure 10.9 Pressure and yield correlation.

10.4 Designed Experiments

If you want to know how your process will run at a temperature of 30 degrees or 60 degrees, you need data in that range. If you want to know how your process will run at 2 PSI or 20 PSI, you need data in that range. If you want a good model of a process that predicts how it will perform at a wide range of temperatures and pressures, then data across the entire range are needed.

Basically, you need to build a box that encompasses the data range that you are interested in looking at further. This is a 2^2 factorial design (shown in Figure 10.10); it consists of two variables at two levels. This model will require four experimental runs to develop. You will want to set your low and high values for the experiments at the extreme edges of what can be run safely. If the reactor vessel is rated for only 15 PSI, the experiments cannot be run at 20 PSI. Thus, your experiment will not tell you how the process will run at 20 PSI.

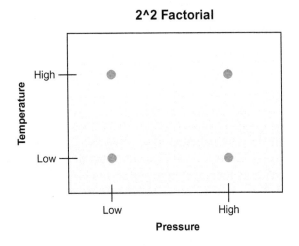

Figure 10.10 Example of a 2^2 (2^2) factorial design.

This is about the simplest designed experiment that can be shown. The only problem with this design is that all of the data are at right angles to each other. This may mask any curvature in the model due to interactions among the variables. To make sure you see this curvature, a center point is added to the design (Figure 10.11).

If you tried to run a process at every possible combination of temperature and pressure, it would be necessary to collect literally hundreds of data points to build the model. By structuring experiments like the ones shown here, a mathematical model can be developed with just four or five experimental runs. If there is concern about the normal variation in the process, the experiments can be repeated in blocks to get two data points at each set of conditions. Designed experiments are a powerful tool for understanding and predicting process performance.

Figure 10.11 Example of a 2^2 (2²) factorial design with centering.

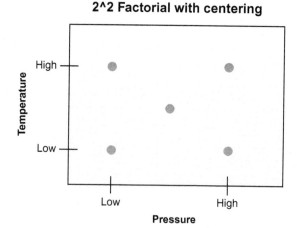

2^2 Factorial with centering

Figure 10.12 Example of form that might be used by process technicians in a designed experiment.

+ = 15 PSI	+ = 65 °C
0 = 10 PSI	0 = 50 °C
− = 5 PSI	− = 35 °C

Run	Rx_pressure	Rx_temperature	Yield
1	+	−	
2	+	+	
3	−	−	
4	−	+	
5	0	0	

When the engineering staff sets up a designed experiment, they will hopefully include process technicians in the discussion about how far the unit can be pushed in all directions. Those discussions may include a form such as the one in Figure 10.12, which is used both to describe the experiment and to provide a place to collect the results.

In Figure 10.12, there are five experimental runs. A plus sign (+) has been used to indicate the high settings and a minus sign (−) to indicate the low settings. Run number 5 is the zero value, or the center point of the design. You can see that this set of runs accounts for each combination of high and low values for both pressure and temperature.

Processes are often complex, with many input variables. There may be readings and controllers for multiple temperatures at multiple locations in the process, plus the various flow rates, pressures, and other readings. To make matters even more interesting, there is often more than one output (response) variable being monitored. For example, the team might desire a model that predicts yield, purity, and some major impurity.

The objective of the experiment may be to define a model that provides the best process conditions to meet all three desired outputs. An input sheet for an experiment with four major input variables and three outputs might look like the one shown in Table 10.2. The rest of the information that is needed is what the variables are, where they are measured and controlled, and what constitutes a high value and low value for each variable.

In the discussion of simple regressions, it was shown how to use the correlation coefficient to gauge how strong the correlation was. In that case, it was easy to draw a line that represented the relationship between the single input variable and the output variable. When running experiments that have four, five, six, or even more input variables, it is impossible to draw

Table 10.2 Sample Input Sheet for a Designed Experiment

Run	Var-1	Var-2	Var-3	Var-4	Response-1	Response-2	Response-3
1	−	−	−	−			
2	+	−	−	+			
3	−	+	−	+			
4	+	+	−	−			
5	−	−	+	+			
6	+	−	+	−			
7	−	+	+	−			
8	+	+	+	+			

the relationship on a graph. Instead, the result of the analysis is a two-dimensional graph that plots the actual values obtained versus the predicted values obtained. If this graph makes a straight line, there is a good correlation. If this graph makes a shotgun pattern, there is not a good correlation.

Earlier, 2^2 factorial experiments were mentioned as being two variables at two levels. A 3^2 factorial is three variables at two levels each. Sometimes you have reason to design experiments that are conducted at more than two levels, so you may have 2^3 or 3^3 experiments. Other types of experimental designs being conducted may include the following:

- Full factorial designs
- Fractional factorial designs
- Central composite designs
- Equiradial designs
- Box-Behnken designs
- Mixture designs

The only point in mentioning these is to give you an appreciation for the complexity of the topic and the diversity of what you may be exposed to later. There is no way you can become an expert in the field of experimental design in one short chapter, but it is important to appreciate your role in the success of the experimental runs.

10.5 Role of the Process Technician

What is expected of the process technician during the execution of experimental designs? The process technician may be expected to do the following:

- Monitor process conditions during the run.
- Take thorough notes about what is observed.
- Gather the requested data.
- Report all of the preceding information accurately and precisely.

In our experience, process technicians have done a great job of monitoring process conditions and gathering the requested data. However, because they did not understand the nature and value of designed experiments, technicians often did not capture notes about observations during experiments that they could pass on to the rest of the staff. Process technicians can be a valuable resource in indicating which data measurement points are nonworking, inaccurate, or unreliable.

Recall our discussion of dual correlation. If a process technician had not mentioned that there was a raw material change in the middle of the experimentation, a wrong conclusion would have been drawn. Sometimes observations will not be related to an experiment.

Sometimes they will. The only way to be sure you have captured everything of value is to *take complete notes and share the information.*

A failure that is seen occasionally is that information gets reported, but not as accurately and precisely as needed. In many organizations, the salaried and hourly staffs work together well, but in a few organizations, these two groups have trouble establishing a high level of trust and understanding.

In those cases, a scenario might arise in which the shift process technician is instructed to run the unit at 50 degrees and 10 PSI, but due to circumstances outside his or her control, this person was able to achieve only 48 degrees and 9.5 PSI. It is not a major failure that the exact experimental conditions were not met. However, it would be a problem if the process technician wanted to avoid a confrontation and inaccurately reported that the process ran at 50 degrees and 10 PSI as requested, with a yield of Y. If these false results are put into statistics software to generate surface plots and other analyses, the analyses would not work out as well as expected because the model was assuming the results were due to 50 degrees when the process really operated at 48 degrees. The model was also assuming the results were due to 10 PSI, but in reality it operated at 9.5 PSI.

The output from multiple regression analysis of data collected from designed experiments is only as good as the data that are put into the system. The conditions at which you are instructed to run are desired conditions. If you cannot reach them for any reason, *document what is actually achieved* so that the model will still be useful. Night shift workers should communicate any issues to the day staff so that they can have the option of making modifications to maintain symmetry in the design.

Roman Catapult Example

Instead of discussing a large process to simulate a designed experiment, we are going to use a catapult (Figure 10.13). The output of this process is the launching of a ball across the room.

Figure 10.13 Catapult.

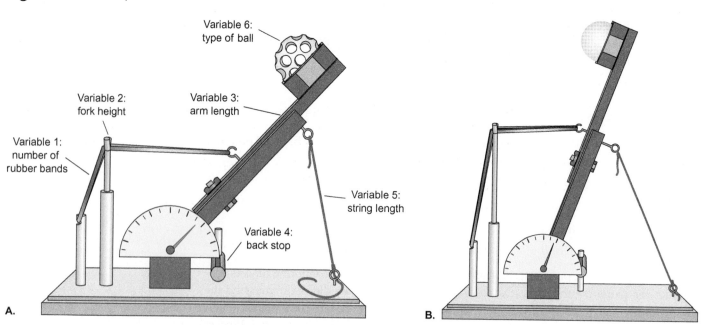

There are at least two measurable outputs from this process: how far the ball travels and how high the ball travels. We can easily measure the travel distance with a tape measure and the height by launching the ball over a barrier. Can you think of another possible process output?

In medieval times, the catapulted objects were required to travel the distance to the target and to clear the top of the castle wall. The process had to be modeled with multiple output variables depending on where the catapult was in relation to the castle wall.

There are several input variables for this process as well. We can put a varying number of rubber bands on the catapult. Different brands and different sizes of rubber bands can be used. Just as in processes, raw materials are not always constant and consistent.

Another input variable is the height of the fulcrum. We have various fulcrum posts and each one is adjustable. Next, we have the length of the catapult arm. The arm may include a measurement system; otherwise, we will have to measure the distance from the pivot point to the ball holder.

The next two variables are the start angle and stop angle of the catapult. Our process comes equipped with a backstop that allows us to start the catapult at an incline. It also comes with a string that can adjust how far the catapult will move forward. With a protractor, we can measure the angles at which to start and stop the process.

Finally, there is the ball itself. We can use ping pong balls, plastic golf balls, whiffle balls, or just about anything that will fit in the ball holder. Each different ball will have a different formula for how to make it fly a certain distance and obtain a certain height.

During one of the first exercises with this example apparatus, one team of students discovered that when they randomly placed their ball in the holder and launched it, they received the same basic result ± 3 inches. When they took care to place the ball in the holder so that the seam was oriented exactly the same way each time, the accuracy remained the same, but the precision improved to ± 1.5 inches. This was a classic case of the process technician (the students) discovering a source of variability that was previously unknown or at least undocumented by the engineering staff (the instructor).

Challenges that can be added to this experiment are hitting a certain target at a distance, hitting an even farther target, and clearing an obstacle between the catapult and the target.

Summary

Designed experiments are a powerful statistical tool for gathering minimum amounts of data that provide maximum amounts of useful information. Designing experiments with purpose instead of collecting random data ensures that they will cover as large a range as possible for the input variables in order to make the model as useful as possible.

The data from designed experiments are used in a multiple regression analysis to generate mathematical models of a process's performance. A process performance model enables companies to customize product offerings to their customers when needed, as well as to make quick adjustments to the process to keep it running on spec and consistently.

Even without the technical knowledge of how to perform the regression analyses, you can determine the relative strength of the model simply by looking at the R^2 values for the model. For simple regressions, you can tell the strength of the model and the direction of the correlation by examining the correlation coefficient.

Designed experiments can generate data that will guide in modeling a process. They can supply information about single or multiple variables and show how variables interact.

The role of the process technician is to monitor the process during the experimentation, take notes about any changes in the process during the experimentation, gather the requested data, and report all information accurately and precisely. Any inaccuracies or estimating of the input or output readings may make the data sheet look good but will defeat the purpose of generating a useful process model.

Checking Your Knowledge

1. Define the following key terms:
   a. Correlation
   b. Performance model
   c. Regression

2. (*True or False*) Modeling predicts how a process will perform based on the input variables.

3. Modeling can be used to predict the performance of which set of data?
   a. The range of data that was used to calculate the model
   b. Any data that the process can handle
   c. A broad range of data
   d. Anywhere between 100 PSI and 110 PSI

4. The model for a simple correlation is often written as:
   a. $Y = x(f)$
   b. $Y = f(v, w, x, z)$
   c. $Y = f(x)$
   d. $Y = mx + b$

5. Select the correct description of the image below:

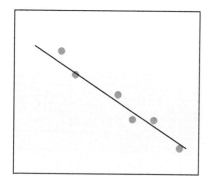

   a. This is a positive correlation.
   b. This is a negative correlation.
   c. This is a zero correlation.
   d. This is a hyperbolic correlation.

6. Select the correct description of the image below:

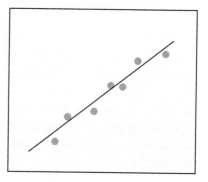

   a. This is a positive correlation.
   b. This is a negative correlation.
   c. This is a zero correlation.
   d. This is a hyperbolic correlation.

7. Select the correct description of the image below:

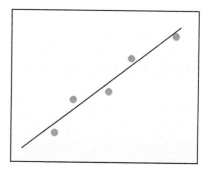

   a. This is a moderate correlation.
   b. This is a strong correlation.
   c. This is a zero correlation.
   d. This is a correlation coefficient.

8. Select the correct description of the image below:

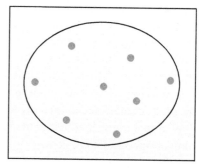

   a. This is a moderate correlation.
   b. This is a strong correlation.
   c. This is a zero correlation.
   d. This is a correlation coefficient.

9. Select the correct description of the image below:

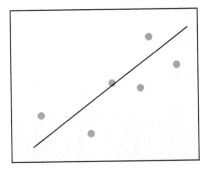

   a. This is a weak correlation.
   b. This is a strong correlation.
   c. This is a zero correlation.
   d. This is a correlation coefficient.

10. It is ideal to achieve a correlation coefficient of at least:
   a. 0.10
   b. 0.30
   c. 0.50
   d. 0.75

11. (*True or False*) The lower the R-sq, the better the model describes the process.

12. A model that consists of two variables at two levels will require how many designed experiments?
   a. 1
   b. 2
   c. 3
   d. 4

13. The R^2 value is best defined as:
   a. the square of the correlation coefficient.
   b. the percentage of the variability in the data explained by the model.
   c. the percentage of the variability in the model that is explained by the data.
   d. the desired target value for all designed experiments.

14. (*True or False*) Correlation does not equal causation.

15. (*True or False*) If you cannot achieve the targeted conditions, do the best you can and write down that you hit them exactly.

16. (*True or False*) Process technicians play a key role in the success of designed experiments.

17. (*True or False*) Only one output variable (response) can be studied at a time in designed experiments.

18. (*True or False*) A perfect correlation has a correlation coefficient of ZERO, which tells you that there were ZERO deficiencies in the model.

19. (*True or False*) Taking notes about what you observe in the process during the experimental runs is critical to the success of designed experiments.

NOTE: Answers to Checking Your Knowledge questions are in the Appendix.

Student Activities

1. The instructor will divide the class into teams to demonstrate the use of the catapult as a designed experiment exercise.
   Here is a potential design for your catapult experiment. What values would you use for the high and low values for each variable?

Experiment Number	Arm Length	Fulcrum Height	Start Angle	Stop Angle	Distance
1	–	–	–	–	
2	+	–	–	+	
3	–	+	–	+	
4	+	+	–	–	

Experiment Number	Arm Length	Fulcrum Height	Start Angle	Stop Angle	Distance
5	0	0	0	0	
6	–	–	+	+	
7	+	–	+	–	
8	–	+	+	–	
9	+	+	+	+	

2. Imagine that you have a barbeque pit. Identify the key input and output characteristics that you might consider including in a designed experiment to model the performance of the pit.

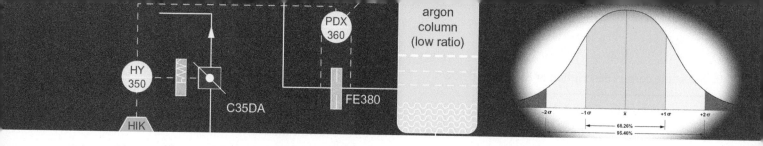

Chapter 11
Other Basic Quality Tools

"You cannot mandate productivity; you must provide the tools to let people become their best."

~ Steve Jobs

 ## Objectives

Upon completion of this chapter you will be able to:

11.1 Describe the check sheet tool and how it is used. (NAPTA Quality, Group Problem Solving 1*) p. 149

11.2 Explain the concept of brainstorming, the rules associated with it, and the circumstances in which brainstorming would be used in an industrial setting. (NAPTA Quality, Group Problem Solving 1) p. 150

11.3 Explain the concept behind the cause-and-effect (fishbone) diagram and apply that concept when capturing the results of a brainstorming session. (NAPTA Quality, Group Problem Solving 1) p. 152

11.4 Explain the use and benefits of work process flowcharts, trend charts, scatter plots, and Pareto charts. (NAPTA Quality, Group Problem Solving 1) p. 152

Key Terms

Brainstorming—a group exercise designed to solicit a large number of ideas in a short amount of time, **p. 150**

Fishbone diagram—a graphic presentation device whereby ideas (or causes) are grouped into common categories such as manpower, methods, machines, and materials. Also known as an *Ishikawa* or *cause-and-effect diagram*, **p. 152**

*North American Process Technology Alliance (NAPTA) developed curriculum to ensure that Process Technology courses will produce knowledgeable graduates to become entry-level employees in process technology. Objectives from that curriculum are named here in abbreviated form. For example, "(NAPTA Quality, Group Problem Solving 1)" means that this chapter's objective 1 relates to objective 1 of the NAPTA curriculum about group problem solving.

Flowcharts—documents that represent a sequence of operations schematically or visually, **p. 152**
Pareto charts—graphics that rank causes from most to least significant; they represent the 80-20 rule, which says that most undesired effects come from relatively few causes, **p. 156**
Trend charts—graphic representations of a direction in a set of statistical data at a particular time. Also, a computer program which allows process variables to be monitored for a set period of time, **p. 154**

11.1 Introduction

This chapter takes a look at some of the basic quality tools that are included in most texts on quality improvement (Figure 11.1).

Figure 11.1 Some basic quality tools.
CREDIT: Courtesy of Willie L. Myles.

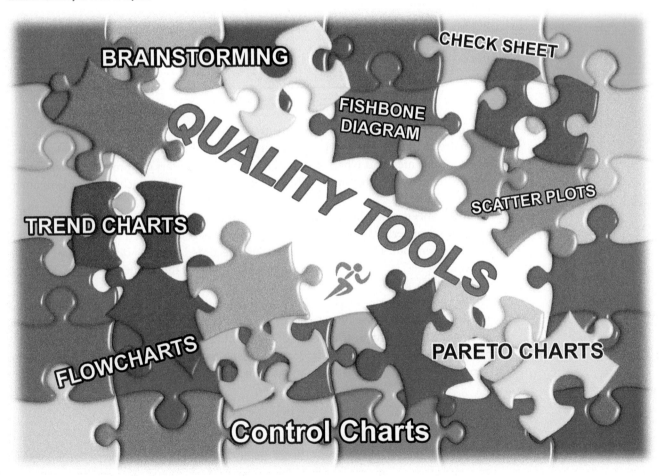

The basic quality tools outlined in this chapter include check sheets, brainstorming, cause-and-effect (fishbone or Ishikawa) diagrams, work process flowcharts, trend charts, scatter plots, and Pareto charts. In Chapter 13, *Variance and Operating Consistency*, we will discuss a very useful tool often included as a "basic quality tool," the histogram.

Each one of these tools plays a part in achieving the overall goal of improving processes so that products and services to the customer also improve. If you satisfy customers, they will continue to buy from your company preferentially, so your company will make money and stay in business.

Check Sheets

The check sheet (Figure 11.2) is a form used to collect data in real time where the data are generated. The data it captures can be quantitative or qualitative. When the information is quantitative, the check sheet is sometimes called a tally sheet.

Figure 11.2 Sample check sheets.

Check Sheet for Calibrating Valve Position

Valve position at 50%	Monday	Tuesday	Wednesday	Thursday	Friday
Test 1	45	50	45	45	55
Retest	50		50	50	50
Test 2	50	55	50	45	50
Retest		50		50	
Test 3	55	50	50	50	
Retest	50			50	

Check Sheet for Instances of Unscheduled Shutdown in First and Second Quarters

Defects in Boardmill Product	January	February	March	April	May	June
Unit 1	///	//			/	
Unit 2	////		//	/		/
Unit 3	//	//		/		
Unit 4	/	//	/			/
Total	10	6	3	2	1	2

Check sheets are part of operations that help to make sure process control devices and instrumentation are working properly. Check sheet information includes how a device is calibrated and when it is checked to see if it is working properly. Check sheets help ensure correct performance. They are generally used by a technician or instrumentation specialist.

Check sheets say what is supposed to happen and can be used as a reference. For example, a drop test may be done on a weigh scale that is supposed to measure x number of pounds per hour. The actual measurement is recorded, and if it does not meet requirements, the speed is adjusted, a second test is done, and the results are entered into the check sheet.

The check sheet can be used as a periodic calibration tool. For example, it can be used to ensure that a valve goes to the correct position. If it is set to 50 percent, it should go to that position rather than to 45 percent or 55 percent. In a test, the actual data about the valve position are recorded. If it is not the correct measurement, the valve is adjusted, and the test is repeated. The second measurement is also recorded in the check sheet.

11.2 Brainstorming

Brainstorming   a group exercise designed to solicit a large number of ideas in a short amount of time.

Whether you are trying to generate a list of quality opportunities to explore, a list of potential root causes of a problem, or a list of possible solutions to a validated root cause, **brainstorming** is a tool to help produce many ideas to examine. Keep in mind that these options are just the opinions of the participants, and data will be needed to determine which option(s) to pursue further. Most quality texts cover some variant of brainstorming.

Brainstorming (Figure 11.3) can be conducted in a structured or an unstructured manner. It can be accomplished in small or larger groups. In some teams, the exercise can be accomplished as a group exercise. Other teams will benefit from having one facilitator lead the exercise. Generally, brainstorming exercises that are part of an improvement team or a root cause analysis team have specific instructions and guidelines. The following description is one example of how the activity might take place.

Figure 11.3 A. and **B.** Brainstorming is a valuable process for coming up with options and potential solutions.
CREDIT: **A.** Sushiman/Shutterstock. **B.** Mint Images Limited/Alamy Stock Photo.

A.

B.

Steps for brainstorming:

1. Assemble a group of three to nine participants. Include some subject matter experts as well as some participants who are not experts in the subject at hand.

2. Have the facilitator review the process to be followed, the rules for brainstorming, and the problem statement.

3. Clear a section of the conference room or control room wall.

4. Provide sticky notes and pens to each participant.

5. Set a 5-minute time limit with no talking. Have each participant write down as many ideas as possible, perhaps suggesting a minimum of 10 ideas per person.

6. At the end of the time, have each participant read his or her ideas and post them on the wall for everyone to see. Do not allow any discussion or criticism of the ideas; just voice them and post them.

7. Group together any ideas that are duplicates as each participant reads his or her ideas aloud.

8. Give everyone a fresh set of sticky notes once all the ideas from round one have been posted and instruct the group to generate 10 more ideas in the next 5 minutes. It is perfectly okay to build on ideas that have already been posted.

9. Have the participants read their ideas aloud one at a time and post them on the wall.

10. Run through the process again if everyone was able to quickly come up with 10 ideas. Keep repeating the process until the participants have no further ideas. (Pushing the process this way is sometimes called "brainstorming to rebellion.")

It is not uncommon for a group of 5 people to generate 100 ideas in less than an hour using this approach.

Certain guidelines need to be followed for the brainstorming process to work optimally. Here are a few of those guidelines:

- Strive for quantity.
- Accept every idea as a potentially good idea.
- Encourage "out of the box" thinking.
- Do not allow discussion—just post the answers as given.

- Do not allow judgment or criticism of others' ideas.
- Encourage participants to build upon ideas.
- Keep the process moving quickly.

11.3 Cause-and-Effect (Fishbone or Ishikawa) Diagrams

Fishbone diagram a graphic presentation device whereby ideas (or causes) are grouped into common categories such as manpower, methods, machines, and materials. Also known as an *Ishikawa* or *cause-and-effect diagram.*

One way of capturing the results of your brainstorming activities is to organize the ideas on a **fishbone diagram** (shown in Figure 11.4), sometimes referred to as an *Ishikawa diagram* after Dr. Kaoru Ishikawa, who first proposed it.

Figure 11.4 Simple example of a fishbone diagram.

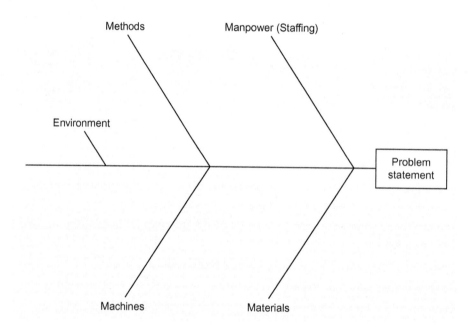

Imagine that you were looking for a small set of categories that would capture every idea generated in a brainstorming session. Dr. Ishikawa suggested that causes of a problem were related to *manpower (staffing or personnel), methods, machines,* or *materials.* As this tool developed, the category of *environment* was added as a way of capturing causes outside of human control, such as hurricanes and earthquakes.

Let us explore an example. A business unit held a brainstorming session during which dozens of ideas were proposed regarding what was causing unplanned shutdowns in the unit. Some of the ideas were the catalyst being used and the product type being produced. Other ideas were related to the utilities used in the manufacture of the product. All of these ideas were grouped under the materials section. They classified some shutdowns as related to quality problems, regulatory problems, or supply problems—these were also materials issues. Some shutdowns were due to the specific machinery. Others were related to human error and further subcategorized by type, position, and the location where people worked. A complete version of this classification exercise was shown in Figure 7.6.

11.4 Other Important Assessment Tools
Flowcharts

Flowcharts documents that represent a sequence of operations schematically or visually.

Flowcharts are an excellent method for ensuring everyone possesses the same understanding of a given issue. The purpose of a flowchart is to represent a process or a portion of a process visually.

Take four shift workers, put them in a conference room with the staff person overseeing the operation, and ask them to draw a flowchart of any given work process. Each person in the room "knows" how to do the job, yet each one may approach the job from a slightly different perspective. If everyone in the room is allowed to contribute his or her perspective, then not only will the resulting flowchart be more comprehensive, but also it may shed some valuable light on where improvement opportunities exist.

To illustrate the point, Figure 11.5 shows an example of a high-level flowchart for changing a flat tire. Diamonds are used to represent questions, or decision blocks. Rectangles are used for process steps. If you were to consult enough different sources, you would find there are dozens of symbols that you can use to make flowcharts specific. For example, circles are often used as pointers to connect portions of flowcharts that cannot fit on a single page. However, most flowcharts can use just a few symbols and still meet the basic needs of the users.

How to Change a Flat Tire

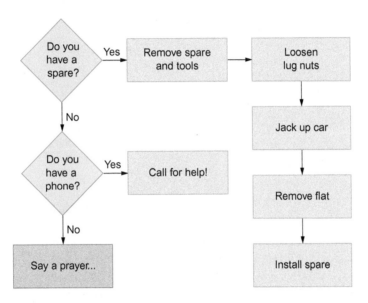

Figure 11.5 Example of flow chart.

Here is another example. First, start with the overall process for how to make isopropyl alcohol (shown in Figure 11.6). For each of these large process steps, there are detailed work process steps.

Process Flow for Manufacture of Isopropyl Alcohol

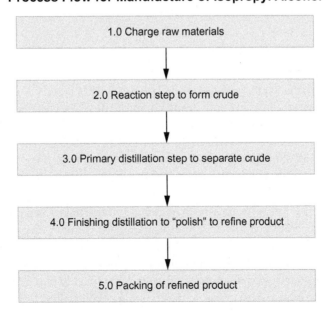

Figure 11.6 Fictional high-level process flow for isopropyl alcohol.

You may be in a position to benefit from seeing the entire process in broad terms. You may be given the job of charging the raw materials, in which case you would benefit more from the detailed work process steps (shown in Figure 11.7).

Figure 11.7 Example of detailed work process steps.

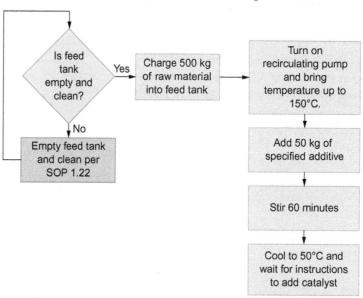

Isopropyl Alcohol Subprocess: 1.0 Charge Raw Materials

Another benefit of flowcharts is that they provide a framework for analyzing the process. In this fictitious example, the technical staff, maintenance staff, and operations staff were assembled to discuss unplanned events. The goal was to discover the reason for the recent downtime and to reduce these events to manageable levels.

With this flowchart on the wall in front of the team, each of the unplanned events of the last 12 months was discussed. Nearly every person in the room had experienced one unplanned event, but no one had been there for all of them. By marking which process steps were involved in the historical unplanned events, they were able to determine that over 80 percent of their problems were related to the "bring temperature up to 150 degrees Celsius" step. In this way, the flowchart can become a simple data collection and data analysis tool, allowing process technicians to see where the problems are occurring so they know where to direct their improvement efforts.

Trend Charts

Because Chapter 12, *Data Collection and Representative Sampling*, Chapter 14, *Variables Control Charts and Interpretation*, and Chapter 15, *Attributes Control Charts and Interpretation* will cover control charts in depth, **trend charts** are described briefly here. Trend charts are simply plots of data against time to check for patterns.

Consider the case of the isopropyl alcohol plant. The operators there check the moisture content of the final product stream every 2 hours. Take a quick glance at the trend chart in Figure 11.8. As you can see, there are no obvious signs that the process is trending up or down. No trends can be determined without further analysis.

If this were an issue of ongoing concern, then you would use these data to establish control limits and then monitor this parameter continuously so that the control chart would alert you to out-of-ordinary performance. The benefit of trend charts is that they are quick and easy to plot—no statistics. If you do not know what you are looking for or how important the various parameters are, then you can quickly plot many variables and visually check to see which ones appear to deserve more exploration.

Trend charts graphic representations of a direction in a set of statistical data at a particular time. Also, a computer program which allows process variables to be monitored for a set period of time.

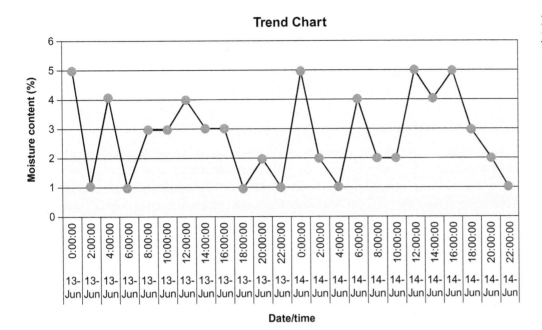

Figure 11.8 Example of a trend chart.

Trend charts are one form of a time series chart, meaning the data are plotted against increments of time. They show how the process changes over time. They are different from histograms, which are snapshots of how a process behaved during a specified time frame.

The data on a trend chart can tell you how your process behaved from one point to another. Each point has a value. Each one tells you something different. Never underestimate the value of putting data up on the wall in graphic format and just studying it for a while. You never know what you might see when you take the time to look.

Scatter Plots

Scatter plots visually indicate two parameters in relation to each other. For example, you could plot the same quality parameter that interested you before, like moisture content, against the temperature readout in the distillation column. Instead of telling how the moisture changed with time, the plot would tell you how the moisture changed with column temperature.

In the example of a scatter plot shown in Figure 11.9, can you predict at what temperature the lowest moisture content is achieved? Hopefully you said no. In this example, you

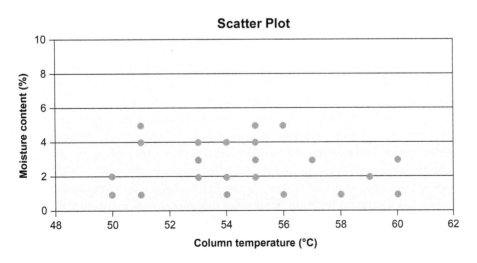

Figure 11.9 Example of a scatter plot.

can see that moisture contents of 1 percent are achieved at 50 degrees and 60 degrees and at many temperatures in between. These temperatures also correlated with higher moisture contents. Here it is enough to know that scatter plots are graphs of two variables or parameters plotted against each other. Review Chapter 10, *Group Problem Solving—Designed Experiments,* to recall correlations in more detail.

Pareto Charts

Pareto charts graphics that rank causes from most to least significant; they represent the 80-20 rule, which says that most undesired effects come from relatively few causes.

Pareto charts are named after a prominent Italian economist named Vilfredo Pareto. He noted that 80 percent of the land was owned by only 20 percent of the people. He called these 20 percent "the vital few," people who controlled 80 percent of the wealth. The remaining 80 percent of the people were the "trivial many." This 80 percent, combined, owned only 20 percent of the country's wealth.

This principle is used extensively in the quality world to describe the relationship between problems and causes (Figure 11.10). You can use this concept to prioritize your improvement efforts.

Figure 11.10 Pareto principle.

CREDIT: OnD/Shutterstock.

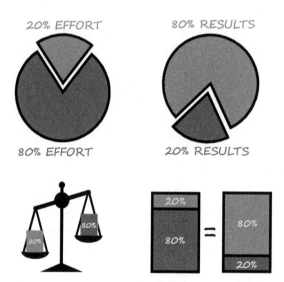

PARETO PRINCIPLE

Your job is to identify which 20 percent of the root causes are responsible for 80 percent of the problems. You can then work to eliminate those causes first, and then reevaluate the data to see where your next priority should lie.

In the Flowcharts section above, the improvement team in the example went through all of the unplanned events and marked which step of the process was active when the event occurred (Figure 11.11). As you have already learned, the "heat to 150 degrees Celsius" step was where most of the problems were occurring.

If you had just taken raw data and charted it instead of making marks on the flowchart, you would have obtained a Pareto chart (shown in Figure 11.12). As you can see, a Pareto chart is a bar chart showing the frequency of occurrence for each of the causes or categories. The categories are sorted by frequency, so the biggest numbers always show up on the left side of the graph. An added feature of most Pareto charts is the cumulative percentage curve, which has a scale on the right-hand side. This shows how many and which categories to group together to reach the 80-percent mark, in this case the Heat and Charge phases.

Isopropyl Alcohol Subprocess: 1.0 Charge Raw Materials

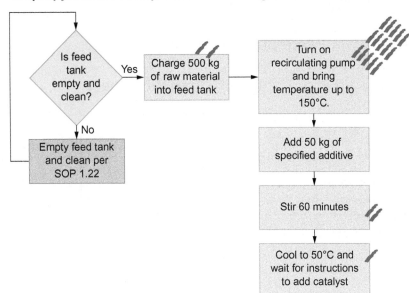

Pareto Chart Unplanned Events
Isopropyl Alcohol Unit 2020/2021 Data

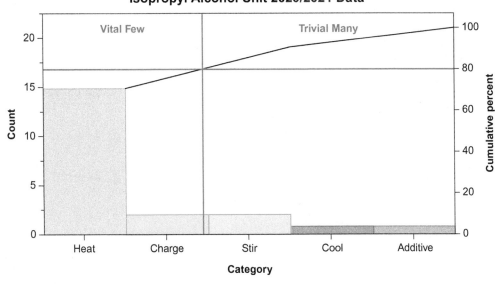

Figure 11.12 Pareto chart of unplanned events tallied in the previous image.

If you presented the data in this format to your management, then you would have no trouble convincing them that you need to improve the heat-up phase of the raw material charging operation. Conclusions are obvious when data are put in the right format.

A Pareto chart is generated as follows:

1. Collect the data.

2. Find the total number of incidents per cause.

3. Add up the total number of incidents.

4. Divide the number of incidents for each cause by the total number of incidents and multiply by 100 to get the percentage for each cause.

5. Sort the results so you can plot the highest numbers to the left.

6. Plot your results.

Figure 11.13 shows a worksheet for steps 1–4, and Figure 11.14 shows the final Pareto chart, using causes of flat tires as an example.

Figure 11.13 Example of a Pareto chart worksheet.

Pareto Chart Worksheet

Flat tire causes

Category	Tally	Subtotal	Percent
Nails	| |	2	6%
Normal wear/tear	| | | | | | | |	8	23%
Cold weather	| |	2	6%
Bullets	|	1	3%
Arrows	|	1	3%
Thorns	| | | | | | | | | | | | | | | | | | |	19	54%
Cause unknown	| |	2	6%

Grand total = 35

Figure 11.14 Example Pareto chart of data from the previous image.

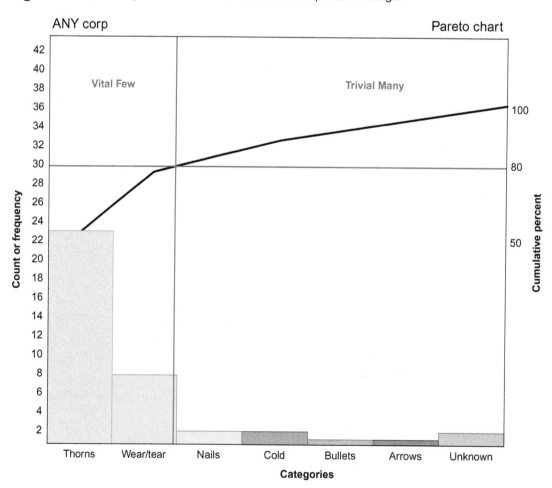

Note that we chose to put the "unknown" category at the right-hand side of the listing of categories, even though it occurred more frequently than the bullets and arrows categories. Because we cannot fix what we do not know, there is no reason to include the unknown in the priority list of things to fix. In Figure 11.14, you can see that two categories, Thorns and Normal Wear and Tear, constitute almost 80 percent of the instances causing the flat tires. If you were to address those two issues, you could eliminate approximately 80 percent of the flat tire problems. Although it is true that we occasionally experience flat tires due to nails, this problem is not the priority.

Summary

This chapter provides a glimpse into the use of a wide variety of quality improvement tools. The overall strategy for quality improvement is one of continuous improvement. You plan out your improvement effort; you do that work using the tools you have learned; you check the results of your work to see whether you accomplished what you set out to accomplish; and finally you act upon the results of your process check. Then you start the planning phase all over again.

The check sheet is a simple tool to capture data in order to ensure that process control devices and instrumentation are working properly. To gather ideas regarding what needs to be worked on or to generate a list of potential root causes to explore further, you might use the brainstorming technique. This is extremely common in the process industries, and process technicians provide a valuable contribution to the effort. A fishbone diagram can be used either to guide the brainstorming effort or as a mechanism to organize the results into categories for further exploration.

Once you have a list of possible root causes, you will be expected to collect data to either validate or invalidate whether each truly is a contributing cause. This is accomplished through flow charts, trend charts, scatter plots, or whatever tool is the most appropriate for the specific situation. If you were looking to see whether a parameter was deteriorating with time, you would use a trend chart. If you were looking to see whether a parameter was statistically stable, you would use a control chart. As you will learn in Chapter 14, *Variables Control Charts and Interpretation* and Chapter 15, *Attributes Control Charts and Interpretation*, the control chart that best fits the need will depend upon the type of distribution underlying the data. If you needed to know how capable your process was of meeting specifications, then you would use a capability graph.

A Pareto chart is the best tool to use when you are determining the relative contribution to an overall problem of the various causes being studied. The Pareto principle states that 80 percent of the problems are due to only 20 percent of the causes. To get the most from your quality improvement, you need to fix the vital few causes that are responsible for the majority of your problems. By knowing and applying the right tools, you help ensure the success of your quality improvement efforts.

Checking Your Knowledge

1. Define the following key terms
   a. Brainstorming
   b. Fishbone diagram
   c. Trend charts
   d. Pareto chart

2. A tally sheet is another name used for a check sheet with which type of information?
   a. Qualitative
   b. Quantitative
   c. Sequential
   d. Periodic

3. (*True or False*) When it is used for periodic calibration, the check sheet only includes the data that reflect the ideal measurement, such as the position a valve should open in the tested conditions.

4. In flowcharts, rectangles are used to represent which of the following?
   a. Process steps
   b. Pointers
   c. Questions
   d. Decision blocks

5. The Pareto principle is also known as:
   a. the 80/20 rule.
   b. the 30/70 rule.
   c. the golden rule.
   d. the 60/40 rule.

6. A group technique for gathering large quantities of ideas in a short time span is:
   a. Pareto charts.
   b. brainstorming.
   c. designed experiments.
   d. group dynamics.

7. Which of these options is a graphic device for grouping causes together?
   a. Pareto chart
   b. Scatter plot
   c. Fishbone diagram
   d. Group dynamics

8. Which of these options is NOT typically included on a fishbone diagram?
   a. Mankind
   b. Methods
   c. Materials
   d. Motors

9. According to the Pareto principle, the 20 percent of the causes responsible for 80 percent of the problems are called:
   a. the trivial many.
   b. the vital few.
   c. the misunderstood.
   d. the Pareto pack.

10. According to the Pareto principle, the 80 percent of the causes responsible for 20 percent of your problems are called:
    a. the trivial many.
    b. the vital few.
    c. the misunderstood.
    d. the Pareto pack.

11. Which tool is used to show changes over time?
    a. Histograms
    b. Scatter plots
    c. Trend charts
    d. Fishbone diagrams

12. Pareto charts have scales on both the left and the right side. What is the scale on the left side used for?
    a. Frequency
    b. Cumulative percentage
    c. Time
    d. Weight

13. Pareto charts have scales on both the left and the right side. What is the scale on the right side used for?
    a. Count
    b. Cumulative percentage
    c. Time
    d. Weight

14. (*True or False*) Trend charts show the relationship of two variables to each other.

15. (*True or False*) Scatter plots are used to show the effect of time on a parameter.

16. (*True or False*) Brainstorming activities are best carried out as a group exercise.

NOTE: Answers to Checking Your Knowledge questions are in the Appendix.

Student Activities

1. The college you are attending has experienced a reduction in enrollment over the last few years and has asked the Process Quality class for assistance.
   a. Conduct a brainstorming session to generate a list of potential reasons for this decline.
   b. Organize the results from your brainstorming exercise into a fishbone diagram.
   c. Review the fishbone diagram as a group. Have each member of your workgroup "vote" for the three potential causes on the fishbone diagram that he or she believes to be the most likely culprits. Place tally marks on the diagram as you go to show which causes are garnering the most votes.
   d. Construct a Pareto chart of your results.

2. Create a flowchart to show a problem you encountered in some previous job. On the left, identify negative factors and where they led, and on the right identify positive factors and where they led. Share your chart with the class.

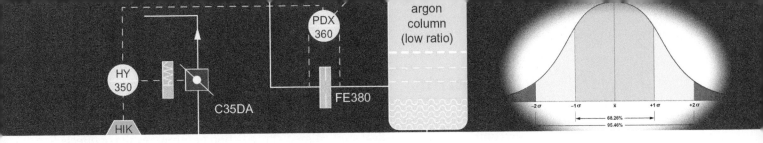

Chapter 12

Data Collection and Representative Sampling

"In God we trust, all others must bring data."

~ Dr. W. Edwards Deming

Objectives

Upon completion of this chapter you will be able to:

12.1 Explain the purpose and uses of plant collection data:

- To describe
- To infer
- To predict. (NAPTA Quality, Data Collection and Control Charts 4, 5*) p. 162

12.2 Explain how data are collected. (NAPTA Quality, Data Collection and Control Charts 6, 7, 9) p. 163

12.3 Discuss various types of data that process technicians would collect and how they represent the process. (NAPTA Quality, Data Collection and Control Charts 2) p. 164

12.4 Explain how data collection is related to troubleshooting. (NAPTA Quality, Data Collection and Control Charts 3, 4, 5) p. 165

12.5 Discuss sampling and representative samples. (NAPTA Quality, Data Collection and Control Charts 2, 3, 10, 11, 12) p. 166

12.6 Describe the role of the process technician in data collection. (NAPTA Quality, Team Projects 1) p. 168

*North American Process Technology Alliance (NAPTA) developed curriculum to ensure that Process Technology courses will produce knowledgeable graduates to become entry-level employees in process technology. Objectives from that curriculum are named here in abbreviated form. For example, "(NAPTA Quality, Data Collection and Control Charts 4, 5)" means that this chapter's objective 1 relates to objectives 4 and 5 of the NAPTA curriculum on data collection and control charts.

Key Terms

Accuracy—exactness; the quality of being exact, **p. 168**

Analytic study—a statistical study aimed at prediction of future results or effects, that begins with evaluation of the results or effects of a cause-system and ends with probable inferences (implicit predictions) about other results or effects not yet observed, **p. 163**

Confidence level—the level of probability that a sample would represent the population, **p. 168**

Enumerative study—a statistical study aimed at the description of a well-defined population; it is an investigation that begins with one or more random samples from a population and ends with inferences about the properties of other elements that were not (but could have been) sampled, **p. 163**

Margin of error—the number of times (in+/−%) a sample would represent the whole, given a specific confidence level, **p. 168**

Operational definition—a description that defines or gives meaning to a variable by describing how it will be measured, **p. 164**

Precision—a measurement of how closely the sample results can be duplicated, **p. 168**

Representative sample—a subset of a population that seeks to accurately reflect the characteristics of the larger group, **p. 167**

Sample—a representative portion of a material collected for analysis, **p. 166**

Size of population—the number of units which make up the whole, **p. 168**

12.1 Introduction

In this chapter, students are introduced to the concept and process of data collection. As an introduction to data collection, students learn how work processes and products are represented by data, and how those data are used to describe, infer, or predict information. During their exploration of the process of data collection, students are provided with an overview of how (1) to describe the data they will collect, (2) to measure the data they will collect, (3) to identify the source of the data they will collect, and (4) to determine the amount of data to collect.

When we refer to the source of our data, we are talking about where we are going to get the data. We already know that a lot of data in the process industry are collected, recorded, and filed. We call this historical data. Data that are yet to be collected will be collected as the result of planned observations. This set is from collection, samples, and readings. How do process technicians determine if their processes are running correctly and if their products meet customer specifications? They sample and analyze their processes and products, then compare results to the process or customer's specifications.

Purpose and Uses of Collected Data

Data are collected all the time. Every day, process technicians take readings and record the data. Process technicians collect samples that are analyzed (Figure 12.1) and record the results. Process control systems (process computers) collect and record data as the process runs.

Process technicians and engineers use collected and recorded (historical) data in several ways:

• To describe their process or products

• To infer what might be going on with their process or products

• To predict what could be done to improve the processes or products.

Figure 12.1 A. The process technician takes samples of polyethylene beads. **B**. Representative sampling.
CREDIT: **A**. Photographic Services, Shell International Limited. **B**. Monty Rakusen/Cultra/Getty Images.

A.

B.

When we seek to *describe* a process or product, we summarize and report the results of observations and collected data.

When we seek to make *inferences* about a process or product, we draw conclusions about something that *will be* presented or observed based on something that *has already been* presented or observed. When data are collected in order to describe or infer, we refer to the effort as an **enumerative study**. Inferential statistics, such as hypothesis testing and analysis of variance, are used in enumerative studies.

Predictions about a process or product are made by taking collected data, which correlate to past performance, and using it to forecast future performance. When data are collected in order to predict, we refer to the effort as an **analytic study**. Control charts and related methods, such as the interpretation and use of the results of an experiment, are used in analytic studies.

Enumerative study   a statistical study aimed at the description of a well-defined population; it is an investigation that begins with one or more random samples from a population and ends with inferences about the properties of other elements that were not (but could have been) sampled.

Analytic study   a statistical study aimed at prediction of future results or effects, that begins with evaluation of the results or effects of a cause-system and ends with probable inferences (implicit predictions) about other results or effects not yet observed.

12.2 The Process of Data Collection

Process engineers and quality managers usually establish data collection procedures for the process technician when the data being collected are for the engineer's or the manager's use. However, if a process technician is working on a team that is responsible for designing its own experiment or study, then the process technician must understand how to plan the process of data collection.

There are a few decisions that must be made in order to proceed with the process of data collection:

- Determine the data we are studying and how we are going to measure the data.
- Determine the method for collecting the data.
- Determine the source of the data to be collected.
- Determine how much data are necessary.

The Operational Definition

Before either an analytic or enumerative study can take place, a fundamental question must be answered: How do we measure the data we are interested in studying? When we wish to thoroughly describe the data in question by explaining how they are to be measured, we

Operational definition   a description that defines or gives meaning to a variable by describing how it will be measured.

create an **operational definition**. For example, if we are talking about measuring human attributes, we can say that intelligence will be measured by an IQ test score, or that height will be measured in inches or centimeters.

If we are talking about a process, how could we operationally define it? Here are some examples:

- Flow rate will be measured in GPM.
- Temperature will be measured in degrees Fahrenheit or Celsius or Kelvin or Rankine.
- Pressure will be measured in inches of water.

Similarly, the following could be operational definitions about a product:

- Contaminants in a chemical compound will be measured in parts per million.
- Viscosity will be measured in centipoise.
- Color will be measured in AP units.

Methods

There are several methods for collecting data. The type of data you are collecting will determine which method is most appropriate. These include the following:

1. *Observing* would be the best method to use when you need to determine how a task is performed. This is done by using the human senses and perhaps also by using instrumentation. Operators are expected to use many observational skills:
   - Seeing
   - Touching (NOTE: Direct touching is prohibited for safety reasons. However, some elements of touching use the nerve endings commonly associated with touch. An example would be noticing unusually high temperature near a machine that is giving off excess heat.)
   - Hearing
   - Smelling
2. *Questioning* is a second method of collecting data. Questioning generally involves asking about a problem, process, or product from multiple angles, interrogating closely in order to understand it fully.
3. *Investigating* is a third method of collecting data. Investigating is looking into, or searching carefully, for facts, knowledge, or information. Investigating usually involves a combination of observing and questioning.
4. *Sampling* is a fourth methods of collecting data. Sampling means collecting a small representative portion of the population, as described later in this chapter.

12.3 Types of Data Collected to Represent the Process

Data that represent the process or the product will communicate whether the product conforms to specifications or does not conform. They indicate that a quality level has or has not been met. Review of data is how process problems are identified. Data, numbers, and statistics tell the process technician when everything is running well and when something is amiss.

Process technicians collect and/or monitor flow rates, pressures, temperatures, and level measurements as part of their routine data collection (Figure 12.2).

Laboratory technicians will collect data on the following on a regular basis:

- pH
- Particle size

Figure 12.2 **A**. Process technician using a "gas sniffer" to check for leaks. **B**. Process technician wearing protective gear for sampling.

CREDIT: **A**. Photographic Services, Shell International Limited. **B**. Courtesy of BASF North America.

A.

B.

- Specific gravity
- Viscosity
- Color
- Conductivity.

12.4 Troubleshooting and Process Data

Data convey when troubleshooting is needed. Inconsistent data mean poor repeatability within a sample or between samples. This is usually caused by instrument problems or unrepresentative samples.

Irreconcilable data mean that the data points or calculations "don't add up." For example, material balances sometimes do not balance. We know from basic chemistry and physics that matter cannot be created or destroyed, so if there is an imbalance of inputs and outputs, that usually means there is an instrument error. Troubleshooting is then required to determine the source of the error. Keep in mind that no instrument gives perfect measurements, so there will be a tolerance level associated with any balance based on the known quality of the measurements.

Calibration of Instruments

It is critical that instruments are checked regularly to ensure they are working properly (Figure 12.3). Process measurements such as temperature, flow, and pressure are typically calibrated by an instrument technician in the lab prior to installation or in situ (in position) during a turnaround.

Online composition measurements are typically checked periodically using a certified reference substance (a standard) with a known composition. Lab instruments are usually checked daily by the lab technician using known standards to ensure that the lab instruments are functioning properly (i.e., within the known tolerance of the instrument). If a problem is discovered, then the instrument needs to be taken offline and recalibrated using suitable standards.

Figure 12.3 Technicians
checking calibration.

CREDIT: Xmentoys/Shutterstock.

Figure 12.3 Technicians
checking calibration.

CREDIT: Xmentoys/Shutterstock.

12.5 Sampling

Process engineers and/or quality managers will typically establish a sampling scheme for the processes within their chemical plants or refineries (Figure 12.4). These sampling schemes are documented in the facility's standard operating procedures (SOPs). The sampling schemes may be documented in the procedure that tells the process technician how to run the process, or they may be documented in a specific sampling procedure.

Figure 12.4 A. Sample cylinder allows collection of pressurized samples in a transportable container. They can be used with a wide range of options including different end connections and valve styles. **B**. Technician checks a gauge reading after obtaining a sample.

CREDIT: **A**. Photographic Services, Shell International Limited. **B**. Slutsky Maksim/Shutterstock.

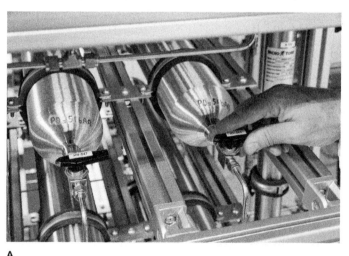

A.

B.

The process of developing a sampling scheme is beyond the scope of this course. However, there are some things about sampling that process technicians should know in order to do their jobs correctly.

We defined a **sample** as a small portion of a population. In the context of the process industries, sampling may refer to collecting a small portion of the product stream at

Sample a representative portion of a material collected for analysis.

whatever point the testing is to be done. Accurate records of the samples are made to document the results found.

Representative Samples

In order to infer meaning to generalize to a larger population, it is absolutely crucial to ensure that it is a **representative sample** of the larger population and that it is documented accurately.

A sample that is not representative can be a serious mistake. Poor samples can result in drawing the wrong conclusions from your data and can potentially cause degradation of product specifications. If you are trying to understand drum-to-drum variability, then it is absolutely critical that the samples be taken from the specified drums.

Samples taken from the wrong location because of convenience can sabotage an improvement effort. Samples taken from a tank that is not well mixed will tell you nothing. In order to understand the quality of the product in the batch tank, you have to know that the contents in the tank are homogeneous (well mixed) so the batch has the same consistency throughout.

Finally, there has to be a sampling methodology in place that ensures your samples are consistent and precise. For example, in Figure 12.5, there is a tank with a recirculating line. Recirculation is a process in which material drawn out of the tank is returned to the tank on a continuous basis per established standard operating procedure (SOP). There is a small sample tap on the discharge of the pump from where you obtain your samples. You should follow the established SOP related to collecting samples. Usually the procedures will state how long to recirculate the tank, how long to purge the sampling line, and how much sample to collect.

Representative sample a subset of a population that seeks to accurately reflect the characteristics of the larger group.

Representative Tank Sampling

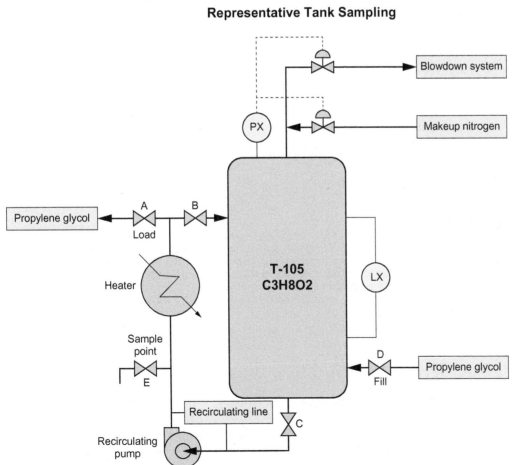

Figure 12.5 Tank with a recirculating line.

CREDIT: Courtesy of Willie L. Myles.

So, what is a representative sample? A representative sample is a sample that can be presented as evidence of the whole. Representative samples must possess the characteristics of accuracy and precision.

Accuracy exactness; the quality of being exact.

- **Accuracy** is how close a measurement is to the accepted or true value of the thing being measured. We can never actually know the true value. Any set of measurements is necessarily an estimate.

Precision a measurement of how closely the sample results can be duplicated.

- **Precision** measures the likelihood that the sample results represent future results.

Did You Know?

Accuracy should never be confused with precision. Precision relates to reproducibility of measurements. Accuracy relates to how close to a true value a measurement is. On a dart board, precision is how closely the darts are grouped. Accuracy is how close they are to the bullseye.

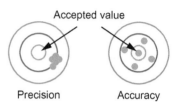

Factors That Affect Sample Size Selection

As was stated previously, process engineers or quality managers typically establish the sample schemes for the facility's processes. These individuals will establish the sample size as well. Process technicians should understand the factors that affect the selection of a sample size so they can understand the rationale and design of the sampling scheme.

When statisticians seek to establish a sample size, they use a calculation that includes the following factors:

Confidence level the level of probability that a sample would represent the population.

- **Confidence level** refers to the statistical chance that a sample will be representative. A statistician tries to achieve confidence intervals between 90 percent and 99 percent. The default value used is usually 95 percent.

Margin of error the number of times (in+/−%) a sample would represent the whole, given a specific confidence level.

- **Margin of error** refers to the statistical chance that a sample would represent the whole, given a specific confidence level.

Size of population the number of units which make up the whole.

- **Size of population** refers to the number of units which can ensure that the sample size is sufficient to represent the whole.

Did You Know?

Incoming Inspection

Unidentified quality problems become more expensive as they move along the process. The goal of an incoming inspection plan is to identify a quality problem (effect) as soon as possible. The first step in developing an incoming inspection plan involves negotiating with the supplier to start an incoming sample program, usually based on the Military Sampling Standard 105E (MIL-STD- 105E) to determine criteria for acceptance and/or rejection of material. This sample program establishes the number of incoming items that must be inspected to ensure a 95% confidence level that the entire shipment will be acceptable.

12.6 Role of the Process Technician in Data Collection

Data collection is a part of a process technician's job. Plants will vary in their job descriptions, but following are some specific tasks technicians may be asked to perform in their professional role:

- Accurately document the data collected, using the appropriate forms and records (Figure 12.6).

Figure 12.6 Process technician documenting data on a digital tablet near the bottling production line in a factory.

CREDIT: wavebreakmedia/Shutterstock.

- *Never falsify data.* Always record what is actually observed, heard, and so on. Some sampling may be performed as part of regulatory compliance. Falsifying such data may lead to civil liability or criminal prosecution.
- Practice discernment—distinguish between what is factual and what is not.
- Use an analytical mindset—keep emotion out of the data collection process. For example, an anticipated solution to a process problem may not be supported by the data. It is important to view the data clearly and objectively. Seeing what you wish to see will not solve a problem in the long run.
- Pay attention to detail—ensure documentation is legible, readings are recorded in the designated place, and so on.
- Do not be complacent—do not behave indifferently when it comes to accuracy and precision.
- Seek clarification whenever something is unclear—if you do not believe the reading on the instrument, ask someone to check it with you.

Summary

Process technicians collect various types of data to describe, infer, and predict. They commonly collect data on flow rates, pressures, temperatures, and level measurements. Proper data collection and sampling are extremely important to process units.

Good sampling is sampling that is representative of the whole process. Bad sampling results in misleading analytical results. Bad samples and poor data collection techniques ultimately result in wasted time and resources.

Process technicians use their senses and defined measuring techniques to interpret the data they have collected. Data are used to troubleshoot issues within the process. Technicians must follow established sampling procedures and practice good data collection techniques to reduce the possibility of generating bad data.

Checking Your Knowledge

1. Define the following key terms:
   a. Accuracy
   b. Analytic study
   c. Confidence level
   d. Enumerative study
   e. Margin of error
   f. Operational definition
   g. Precision
   h. Representative sample
   i. Sample
   j. Size of population

2. (*True or False*) Data that have been collected, recorded, and filed are known as historical data.

3. Which of the following was NOT listed in the chapter as a type of data a process technician may collect on a regular basis?
   a. Temperature
   b. Particle size
   c. Pressure
   d. Level

4. (*True or False*) Data collected from a process will always indicate that a quality level has been met.

5. The study which uses collected data in order to predict future results or effects is called:
   a. Enumerative study
   b. Inferences study
   c. Analytic study
   d. Prediction study

6. (*True or False*) A process control system collects, analyzes, and records samples.

7. Which study is defined as a statistical study aimed at a well-defined population?
   a. Inferences study
   b. Probable study
   c. Prediction study
   d. Enumerative study

8. Which method does the process technician use when determining how a task is performed?
   a. Questioning
   b. Investigating
   c. Sampling
   d. Observing

9. When establishing data collection methods, what would the process technician include? (Select all that apply.)
   a. Source of the data
   b. What data are unnecessary
   c. How the data will be measured
   d. Method for collecting data

10. (*True or False*) If a problem in an instrument is discovered, the instrument will be taken offline and recalibrated.

11. Troubleshooting is needed when the process technician finds what?
    a. Consistent data
    b. Reconcilable data
    c. Sampling data
    d. Irreconcilable data

12. What is the process in which material drawn out of the tank is returned to the tank on a continuous basis?
    a. Observing
    b. Recirculation
    c. Sampling
    d. Representative sampling

13. A calculation for a sample size would include which variable?
    a. Size of population
    b. Level of accuracy
    c. Representative sample
    d. Small portion of a population

14. (*True or False*) Accuracy is a measurement of how closely the sample results can be duplicated.

15. (*True or False*) A process technician's data collection will never vary between jobs.

16. When collecting data, which actions does the process technician's job include? (Select all that apply.)
    a. Paying attention to detail
    b. Recording what the technician thinks was observed or heard
    c. Using an analytical mindset
    d. Accurately documenting the collected data

17. (*True or False*) A process technician's data collection will never vary from one job to the next.

NOTE: Answers to Checking Your Knowledge questions are in the Appendix.

Student Activities

1. This exercise is to highlight what we mean by operational definition. For example, specific gravity would have X result, and color (APHA) would have Y result. Using the certificate of analysis shown below, complete the operational definition questions that follow the chart.

CERTIFICATE OF ANALYSIS (COA)
PROPYLENE GLYCOL
REAGENT ACS & USP/NF GRADE
KOSHER

Test	Monograph	Specification	Result
Assay (on anhydrous basis)	ACS USP	NLT 99.5%	99.99%
Identification A—infrared absorption	USP	Conforms to reference spectrum	Pass
Identification B—Limit of:	USP		
Diethylene glycol		Diethylene glycol, NMT 0.1% max	< 0.1%
Ethylene glycol		Ethylene glycol, NMT 0.1% max	< 0.1%
Identification C—GC	USP	Conforms to reference chromatogram	Pass
Specific gravity	USP	1.035–1.037 @ 25°C max	1.035
Color (APHA)	ACS	NMT 10 max	1
Acidity	USP	NMT 0.20 mL 0.10N NaOH max	Pass
Water	USP ACS	NMT 0.20% max	0.02%

Approved by: *John Doe*

Quality Assurance Technician

- Specific Gravity result: _____
- Color (APHA) result: _____
- Water result: _____

2. Shore tank T-105 (see graphic below) is made out of 1-inch stainless steel and can hold up to 20 million pounds of propylene glycol. Tank is nitrogen padded and does not contain an internal stirrer or mixer. The producing plant sends product to the tank every three days. Within a 30-day period at least 10 different batches of material are discharged into the tank. Prior to distribution from the tank, process technicians are required to collect a representative sample from the tank.

Arrange the following process activities in proper sequence to obtain a homogeneous (well mixed) sample to ensure tank material has the same consistency.

_____ Obtain sample container (bottle) to collect sample.

_____ Recirculate tank for 24 hours prior to sampling tank. Indicate which valve must be opened and closed in order to place the tank on recirculation.

_____ Sample line flush—flush two line volumes.

_____ Collect a sample off the pump discharge. Indicate which valve must be opened and closed in order to catch a sample.

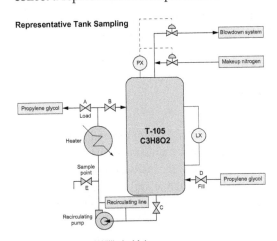

Representative Tank Sampling

CREDIT: Courtesy of Willie L. Myles.

Chapter 13

Variance and Operating Consistency

"Commit yourself to quality from day one. Concentrate on each task, whether trivial or crucial, as if it's the only thing that matters (it usually is). It is better to do nothing at all than to do something badly."

~ MCCORMACK

Objectives

Upon completion of this chapter you will be able to:

13.1 Explain the concept of variability and use a histogram to illustrate the distribution of data. (NAPTA Quality, Variance and Operating Consistency 1*) p. 173

13.2 Calculate measures of central tendency. (NAPTA Quality, Basics of SPC 6, 7) p. 178

13.3 Calculate measures of spread. (NAPTA Quality, Basics of SPC 6, 7) p. 180

13.4 Explain the differences between common cause and special cause, and between a stable process and an unstable process. (NAPTA Quality, Variance and Operating Consistency 2, 3, 4, 5) p. 183

13.5 Discuss the types of data that are represented by normal, Poisson, and binomial distributions. (NAPTA Quality, Control Charts and Data 4, 5) p. 184.

*North American Process Technology Alliance (NAPTA) developed curriculum to ensure that Process Technology courses will produce knowledgeable graduates to become entry-level employees in process technology. Objectives from that curriculum are named here in abbreviated form. For example, "(NAPTA Quality, Variance and Operating Consistency 1) means that this chapter's objective 1 relates to objective 1 of the NAPTA curriculum on variance and operating consistency.

Key Terms

Binomial distribution—distribution of discrete data that are made up of data that have only two choices (e.g., pass/fail), **p. 185**

Central tendency—center or middle of a distribution of data; it is measured as the mean, median, or mode, **p. 178**

Common cause variation—variables in a process that cause the process to vary and are built in and inherent to the process. The variation of the process due to common causes is the variation that is always there when things are running normally, **p. 183**

Continuous data—numbers that can be of any value or fraction of a value, **p. 184**

Discrete data—numbers that are not continuous (e.g., whole numbers). Also known as *variables data*, **p. 184**

Histogram—a type of graphical presentation of data, **p. 174**

Mean—mathematical average of a set of data, **p. 178**

Median—geometric center of a data set, **p. 178**

Mode—number that occurs most frequently in a data set, **p. 178**

Normal distribution—distribution of continuous data that is bell shaped. The data in a normal distribution are distributed such that approximately 68 percent of the data fall within 1 standard deviation of the mean, approximately 95 percent of the data fall within 2 standard deviations of the mean, and approximately 99.7 percent of the data fall within 3 standard deviations of the mean. Also known as a *Gaussian distribution*, **p. 181**

Poisson distribution—distribution of discrete data that consist of counting data. In this case, only whole numbers are present, **p. 185**

Range (R)—the limits, from minimum point to maximum, for any operating condition or specification. Usually used to refer to temperature, pressure, level, specification, flow, or product purity, **p. 180**

σ—lowercase Greek letter sigma; used in mathematics to denote the standard deviation, **p. 180**

Σ—uppercase Greek letter sigma; used in mathematics to indicate a summation, **p. 179**

Special cause variation—event-related items in the process that cause the process to vary outside of its normal operation. These causes are in addition to the common cause variation sources. Also known as *assignable cause variation*, **p. 183**

Spread—the broadness of the distribution of the data set; it is measured as the difference between the highest and lowest values in the data set or by using the standard deviation, **p. 180**

Stable process—state of a process when only common cause variation is present; also stated as the absence of special or assignable causes. A stable process is, by definition, predictable, **p. 183**

Standard deviation—statistical measure of the spread of a set of data, represented by lower case *sigma* (σ), **p. 180**

Unstable process—state of your process when both common cause and special cause variation are present. An unstable process is, by definition, unpredictable, **p. 184**

x-bar—used in mathematics to indicate the mean, **p. 178**

13.1 Introduction

Chapter 1, *Introduction to Process Quality*, defined and discussed the concept of quality. One major component of the definition of quality is consistency. Consistency is uniformity or agreement among things or parts. Customers like consistency. They want a product to look and act the same way every time they buy it. The opposite of consistency is variability.

Variability is observable differences or characteristics between seemingly similar products. Variations in product quality or variation in the service provided is not a good thing from a customer perspective.

Think about the stores you have been to where the service is inconsistent—where you sometimes get great service and other times you get poor service. Now think about those stores where you have consistently had good service every time. Chances are you will be much more inclined to return to stores with consistent service instead of those where service is variable.

Histogram a type of graphical presentation of data.

This chapter explores the concept of variability and proposes some ways to measure and define variation in terms that you can use to improve your processes. The graphic technique that will be employed to represent variability in a process is called a histogram. A **histogram** is a bar chart that shows the distribution of a set of data. Figure 13.1 shows one example of a histogram.

Figure 13.1 Sample histogram.

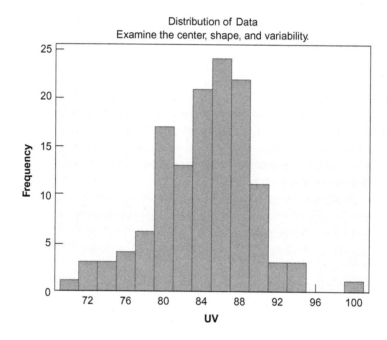

Data distributions can take on many different shapes depending on the type of data that are being collected and the stability of the process generating the data. This chapter explores several different types of data that are important in the processing industries and the types of distributions that best represent those data types.

Regardless of what type of distribution you are dealing with, there are two fundamental ways of describing a distribution using mathematics: the central tendency and the spread.

You will learn how to measure each of these important concepts and how to use these measurements to analyze your processes to determine when the process is stable (exhibiting only common cause variation) versus unstable (exhibiting special cause variation).

The specific branch of mathematics that helps in understanding variation and distributions is called *statistics*. Statistics can be confusing for some, but do not be alarmed. This chapter approaches statistics from the standpoint of making it as simple as possible to use for improving processes. The concepts introduced in this chapter are the foundation for what will be explored in the next few chapters. It is important to take the time to be sure you understand these basic ideas.

This is the first of several chapters on statistics, which is one of the most important pieces of the quality puzzle.

Variability

Everything varies. This is so true that it might be considered a law of nature. If you find something that does not appear to vary, it probably means the variation is so small that you cannot measure it; but it does still vary. Common familiar examples of variability include snowflakes and fingerprints (Figure 13.2).

Figure 13.2 Snowflakes **(A)** and fingerprints **(B)** both have small variations that are not easy to see.
CREDIT: A. Maria.K/Shutterstock. **B**. Rayyy/Shutterstock.

A.   B.

The forces of nature that create snowflakes have never been found to create exactly the same crystalline structure from one snowflake to the next. This can be easily seen when you compare snowflakes created in different temperatures and different places. On the other hand, even snowflakes created at the same place on the same day and at the same time are still different.

There are billions of people in the world, and yet no two fingerprints have ever been found to be the same. When we consider that these people have different parents, different genes, and different environments, we can accept that they will have different fingerprints. Think about siblings who have the same parents, are reared in the same environment, and come from the same gene pool. They still have different fingerprints. Even though identical twins may be so similar that only those closest to them can tell them apart, they are different people with different personalities and different fingerprints.

The products that are sold also vary. Many things contribute to the variability of these products, such as the raw materials that are used. This should not come as a big surprise because one company's raw materials are someone else's product. Also, the process used to convert raw materials into a new product varies from one company to the next.

Think about your family car. You might buy gas from the same station and use that gas in the same car, but your gas mileage probably varies from tank to tank. Some of this could be due to the driving conditions experienced (using one tank for in-town travel and another tank on a highway). But, it could also be due to variation in different loads of gasoline in the station's storage tank.

The most extreme differences would likely be between a full tank used only for highway driving versus a full tank used only for stop-and-go driving in the city. Yet, even two tanks of stop-and-go driving will give two different results depending on the driving conditions.

Even if the discussion is limited to only highway driving and using the exact same stretch of road, the car itself has some inherent (built-in) variability. By the time you start your second trip to demonstrate that the car gets the same mileage every time, the tires are a little more worn and the spark plugs have had more hours of use on them.

The person behind the wheel also plays a role in variation. In this case, the driver can easily affect the results by how he or she operates the vehicle. In the process industries, the process technician is like the driver of the car.

Statistical quality control (SQC) and statistical process control (SPC) are built on the premise that all processes have inherent variation and that this variation does the following:

- Behaves in a predictable manner
- Is stable over time
- Is measurable
- Is random by nature.

Once you accept these truths you can begin your work to improve processes. Your job is to quantify the variation exhibited by the process. Let us illustrate by creating a process, running the process to collect some data, and then plotting the data to demonstrate the journey of quality improvement.

In this example, paper clips are purchased as raw materials. Your plant's process is to break the paper clips in half in order to sell the parts to customers (Figure 13.3).

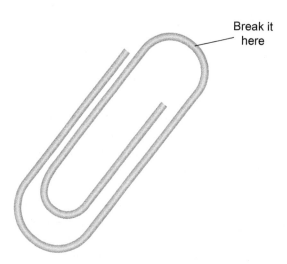

Figure 13.3 Breaking of paper clips.

Workers use the following instructions: Gather 100 paper clips. Notice that one side of the paper clip has two curved portions and the other side has one curved portion. Bend the paper clip back and forth and break the paper clip at the curve on the side that has just one curve. See Figure 13.3. Count the number of bends it takes to break the paper clip.

On the chart in Figure 13.4, color in the box along the bottom of the graph (X-axis) that corresponds to the number of bends required to break the first paper clip. In the case of the example, the first paper clip broke in six bends, so the box above the six is colored.

Break the second paper clip. Record the results on the chart. Repeat this process for the remaining paper clips.

In the experiment, this same process was run for 100 paper clips. The chart seen in Figure 13.5 was then generated. The chart is called a histogram. Histograms are just bar charts showing how frequently each result is obtained from a process and how the data are distributed. These distributions are the foundation for both SQC and SPC.

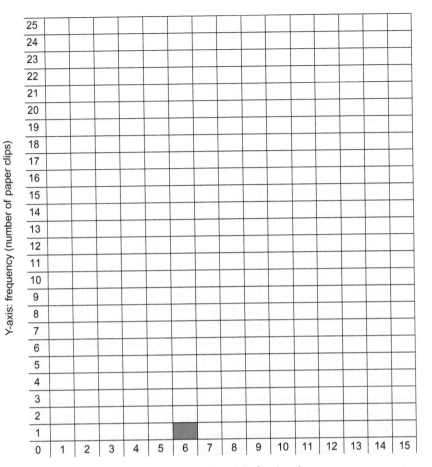

Figure 13.4 Paper clip blank histogram chart.

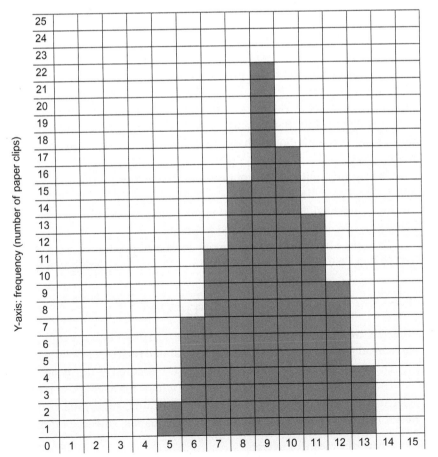

Figure 13.5 Example of paper clip histogram.

If you take samples of data from a process periodically and plot the results in a histogram, assuming the process has not changed, the results would look similar—not exactly the same, but similar.

Your chart will not look exactly like the example in Figure 13.5 nor will it match exactly the charts of others. Some examples of process variations that might contribute to the variability seen in the final histograms include the following:

- Speed of bending might heat up the metal and make the paper clips break faster.
- Degree of bending might vary.
- Bending straight versus bending with a twist would affect results.
- Different people would have different paper clips of different sizes and from different manufacturers.

Variability might be reduced in specific ways:

- Each person could work close to the same speed and bend his or her paper clips the same way.
- There could be just one bender (one process technician) to eliminate differences between one person and another.
- All the paper clips could be drawn from the same box to eliminate the size and composition variables.

Customer specifications will determine the proper procedure. You will have to analyze your process to determine the best way to run it in order to give the customer the desired product consistently. If you are going to train people to operate the process the same way every time, then you are going to need a procedure to use in training. The purpose of documented work procedures is to define the best way of accomplishing a specific task so that everyone can do it the same way every time. This organized approach has several names: *operating discipline*, *standard operating procedures (SOPs)*, *work instructions*, and dozens of others. They all have the same purpose, which is to document the best way to get the work done to ensure a consistent result.

Stories about process technicians who keep personal notebooks in their lockers so they can keep track of personal best practices are legendary. Their actions suggest that A shift is in competition with B shift, which is in competition with C shift, to see who can perform the best. These kinds of work practices are counterproductive to the quality effort.

In reality, A shift, B shift, and C shift are not in competition with each other. Rather, they are all in competition with the other companies that supply the same products to the marketplace.

The various shifts within your company have to work together to make the most saleable product possible. If one shift comes up with a procedure that is better than the one currently documented, it should be shared and the procedure should be documented so that everybody can employ the improved technique.

Consistent operation means a more consistent product. Consistency means customer satisfaction, staying in business, and ultimately protecting your own employment.

13.2 Central Tendency

There are two main measures or statistics used to describe a distribution: central tendency and spread (shown in Figure 13.6). Let us examine each one in a little more detail.

The first measure is **central tendency**; it is measured as the **mean**, **median**, or **mode**.

The measure that most of us are familiar with is the average, or mean value. This is calculated by dividing the sum of the individual measurements by the total number of measurements. In order to write out the formula for this equation, you need to know a couple of terms:

- **x-bar**—the mean, or average
- *x*—the individual measurement

Central tendency center or middle of a distribution of data; it is measured as the mean, median, or mode.

Mean mathematical average of a set of data.

Median geometric center of a data set.

Mode number that occurs most frequently in a data set.

x-bar used in mathematics to indicate the mean.

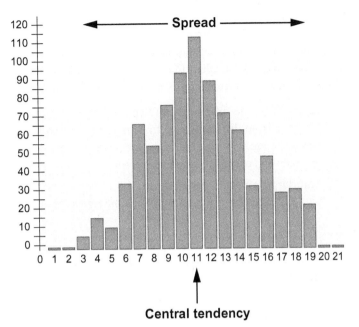

Figure 13.6 Central tendency and spread.

- *n*—the number of measurements
- Σ— the sum, or total.

Σ uppercase Greek letter sigma; used in mathematics to indicate a summation.

The mathematical formula for the mean is shown in Figure 13.7. It is read as x-bar equals the sum of the individual *x*s divided by the number of *x*s (*n*).

Figure 13.7 Mean formula.

$$\overline{X} = \frac{\sum x}{n}$$

Let us take the following five numbers and find the mean: 2, 3, 9, 9, and 7.

Add them up (2 + 3 + 9 + 9 + 7 = 30), then divide by 5 since there are 5 numbers. We know that 30 divided by 5 equals 6, so our mean, or x-bar, is 6.

If you put these same 5 numbers in order and choose the middle number, you conclude that the *median* is 7. The *mode* is 9 because that is the number that occurs most often.

It is not important to know why, but when these three measures of the central tendency are close, it indicates that the data are normally distributed. When these three measures are not close, it indicates that the data are not normally distributed. The next section covers more about normal distributions.

Look again at the histogram of 100 broken paper clips in Figure 13.5. There were 2 results of 5 bends, 7 results of 6 bends, 11 results of 7 bends, and so on. If you add all 100 of the results, then divide by 100, you calculate a mean of 9.23.

By looking at the graph you can determine that the mode is 9 because 9 bends is the number that occurs most frequently. If you sort all 100 numbers in numerical order, you will find that 9 is also the median. Because these numbers are close, we can guess, without any sophisticated statistics, that the histogram is normally distributed.

Spread the broadness of the distribution of the data set; it is measured as the difference between the highest and lowest values in the data set or by using the standard deviation.

Range (R) the limits, from minimum point to maximum, for any operating condition or specification. Usually used to refer to temperature, pressure, level, specification, flow, or product purity

Standard deviation statistical measure of the spread of a set of data. Represented by lower case *sigma* (σ).

σ lowercase Greek letter sigma; used in mathematics to denote the standard deviation.

13.3 Spread

The other main measure of statistics is spread. **Spread** is measured as the difference between the highest and lowest values in the data set. The data spread of the paper clips was 8 because the numbers ranged from 5 to 13. It can also be determined by using the standard deviation.

1. **Range (R)**—the difference between the high and low values in a group of numbers. The data from the 100 paper clips ranged from 5 to 13. Your histogram may have had a similar range or, if you used a different size or type of paper clip or a different technique for bending, could have been different.

2. **Standard deviation**—a measure of the amount of variation or dispersion of a set of values. It is commonly referred to as the *sigma* (σ) of your process.

Did You Know?

This is the lowercase Greek letter *sigma*. It is used in mathematics to denote the standard deviation.

Two negative numbers multiplied together yield a positive number as a result.

$$(-) \times (-) = +$$

The formula for the standard deviation is shown in Figure 13.8. When in the field, you will not have to calculate this number by hand. There are certain tasks that computers are made for, and calculating standard deviations is one such task. You just need to know what it means and how to use it to improve your process.

Figure 13.8 Standard deviation formula.

$$\sigma = \sqrt{\frac{\sum (x - \bar{x})^2}{n - 1}}$$

In order to calculate the standard deviation, follow these steps:

1. Add up the five numbers from the left column of the following chart to get 150, and then divide the total by 5 (*n*) to get the mean or $\bar{x} = 30$. Next, to accomplish the part of the

equation inside the parentheses, subtract this average from each one of the individual values. The results are shown in the right column of the chart:

x	x − x̄	Difference
26	26 − 30 =	−4.00
24	24 − 30 =	−6.00
33	33 − 30 =	3.00
27	27 − 30 =	−3.00
40	40 − 30 =	10.00

2. Square (multiply by itself) each one of these differences:

Difference	Square the Difference
−4.00	16.00
−6.00	36.00
3.00	9.00
−3.00	9.00
10.00	100.00

3. Take the sum of (Σ) these squared differences:

$$16 + 36 + 9 + 9 + 100 = 170$$

4. Divide this sum by $(n - 1)$, where n is the number of pieces of data in the data set. In this example, there are 5 numbers in the data set, so $n - 1 = 4$.

$$170 \div 4 = 42.5$$

5. The final step is to take the square root of 42.5. The answer is the standard deviation $(\sigma) = 6.5$.

When your data fit under a bell-shaped curve, it means your data are normally distributed. This situation is commonly referred to as a **normal distribution** (or *Gaussian distribution*). An example of a normal distribution is shown in Figure 13.9.

Normal distribution distribution of continuous data that is bell shaped. The data in a normal distribution are distributed such that approximately 68 percent of the data fall within 1 standard deviation of the mean, approximately 95 percent of the data fall within 2 standard deviations of the mean, and approximately 99.7 percent of the data fall within 3 standard deviations of the mean. Also known as a *Gaussian distribution*.

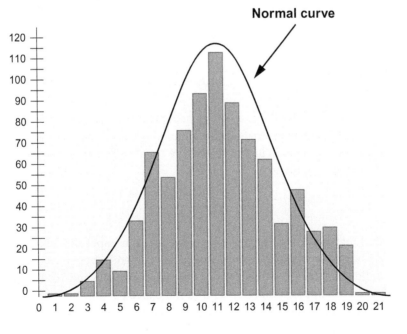

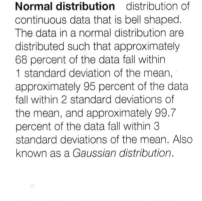

Figure 13.9 Normal distribution.

The standard deviation (Figure 13.10) can be used to determine how much of the data are gathered close to the center of the curve and how much are distributed out toward the edge of the curve. It can even indicate how far from the center the edges should be.

Figure 13.10 Standard
deviation graph.

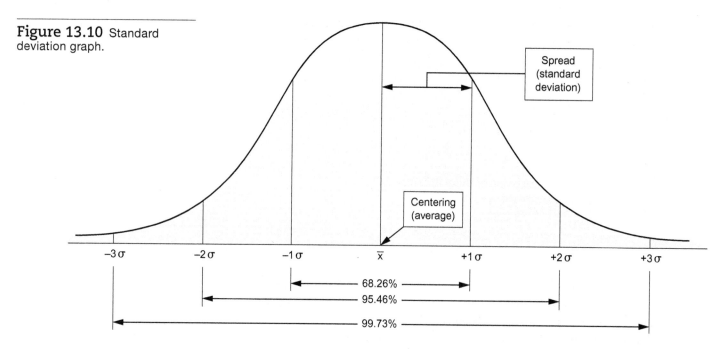

For a normal distribution, approximately 68 percent of the data fall within 1 standard
deviation of the mean, approximately 95 percent of the data fall within 2 standard deviations
of the mean, and approximately 99.7 percent of the data fall within 3 standard deviations of
the mean, as shown in Figure 13.11.

Figure 13.11 Deviation graphs.

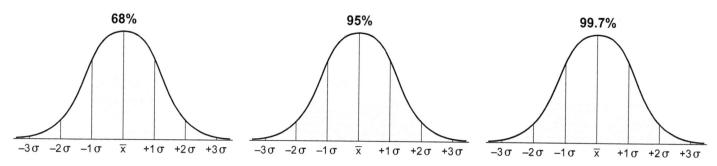

Let us consider the case of our paper clip breaking example. We calculated a mean of
about 9. It so happens that this distribution has a standard deviation of approximately 1.5.
Using the information just provided, we could calculate that 68 percent of our data should
fall within 1.5 units of 9.

$$9 + 1.5 = 10.5$$
$$9 - 1.5 = 7.5$$

We would conclude that 68 percent of the time our paper clips should break between 7.5 and
10.5 bends. We did not capture data in partial bends, but an examination of our histogram
would show that 53 percent of our data fell between 8 and 10, and 77 percent fell between
7 and 11; it appears that our data fit the expected result of 68 percent within 1 standard
deviation of the mean.

Let us go the next level. Ninety-five percent of our data should fall within 2 standard
deviations of the average.

$$9 + (2 \times 1.5) = 12$$
$$9 - (2 \times 1.5) = 6$$

According to our actual data, 94 percent of the paper clips broke between 6 and 12 bends. Finally, we check to see how we did at 3 standard deviations from the mean:

$$9 + (3 \times 1.5) = 13.5$$
$$9 - (3 \times 1.5) = 4.5$$

Our data all ranged from 5 to 13, so 100 percent of our paper clips broke within 3 standard deviations of the mean. We expected 99.7 percent of them to break in that range, so it appears that our actual data fit what we would expect from a normal distribution. There are more accurate tools for determining whether our process is normally distributed, but as a quick check, this method works well.

Understanding the concept of how much of the data falls within a certain sigma range of the mean is critical to the quality improvement effort.

In checking the paper clip example against the expected norms, we found that our data fit. Knowing that the data fit a normal distribution and knowing how to describe that distribution using the mean and standard deviation allows us to predict the future.

Anyone who is into gambling will like these odds. If you bet that the paper clip will break in 7.5 to 10.5 bends, you will win 68 percent of the time. If you bet that the paper clip will break in 6 to 12 bends, you will win 95 percent of the time.

Normal distribution is the desired state. If a process is normally distributed, and nothing abnormal is happening, like a pump breaking down or a line developing a leak, it is possible to predict how a process will perform in the future.

13.4 Cause and Stability
Common Cause *versus* Special Cause

When a process is performing normally and the only variation evident is the random variation that is inherent in the process, you see **common cause variation**.

When only common cause variation is present, expect to see 95 percent of the data within 2 standard deviations of the mean. Expect to see data farther than 2 standard deviations of the mean only 5 percent of the time.

If more than 5 percent of the data are 2 standard deviations away from the mean, it is a good indication that something is wrong. It may not immediately be clear what has changed, but you can be confident that something has changed.

The changes that cause a process to look abnormal might be the shutdown of a pump, a leak in a pipeline, an instrument failure, or an electrical problem, just to name a few. These events are called special causes (or assignable causes) because they cause the process to vary outside of its normal operation. **Special cause variation**, also known as *assignable cause variation*, represents an abnormal variation that can be assigned to a specific event.

Later chapters will examine in more depth the kinds of indications that the data may provide to indicate when something is wrong with a process.

Stable Processes

The process is called stable when it looks the same every time and is considered predictable. A **stable process** is consistent, with the absence of special or assignable causes.

When only common cause variation is present, whatever the process produces today is what it will produce tomorrow, the day after that, and the day after that. This is the desired state.

Given the representation of a process in Figure 13.12A, it is possible to predict what the outcome from this process will be on day 4. This consistency is exactly what the customer desires. Customers want assurance that what they get tomorrow will look and act just like what they got yesterday and last month.

Common cause variation variables in a process that cause the process to vary and are built in and inherent to the process. The variation of the process due to common causes is the variation that is always there when things are running normally.

Special cause variation event-related items in the process that cause the process to vary outside of its normal operation. These causes are in addition to the common cause variation sources. Also known as *assignable cause variation*.

Stable process state of a process when only common cause variation is present; also stated as the absence of special or assignable causes. A stable process is, by definition, predictable.

Unstable process state of your process when both common cause and special cause variation are present. An unstable process is, by definition, unpredictable.

The process is called unstable when it looks different from time to time and is considered unpredictable. An **unstable process** has both common cause and special cause variation. When this situation occurs, it is uncertain whether what the process produces today will be the same thing it will produce tomorrow and every other day.

Given the representation of a process in Figure 13.12B, it is almost impossible to predict what the outcome from this process will be on day 4. From the customer's perspective, your process is unstable, unpredictable, and undesirable. An unstable process may be evidence of special cause or assignable cause variation.

Figure 13.12 A. Stable process. **B**. Unstable process.

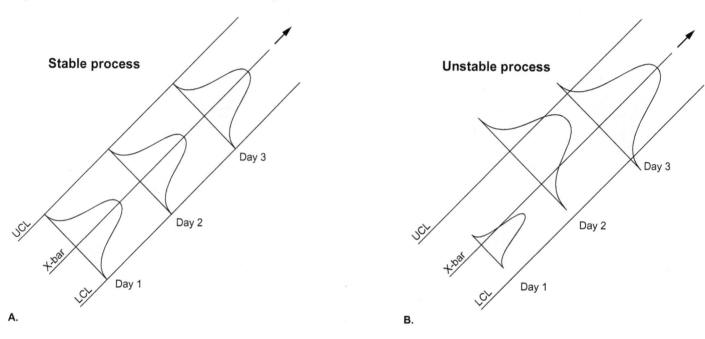

Your customers will not know until they receive and try to use your product how it might behave in their process. This type of performance could cause you to lose business if your competitors are able to produce a stable product.

13.5 Types of Distributions

The discussion so far has been about normal distributions, but there are many other types of distributions. Each one has its own purpose and value. In the processing industries, three types of distributions are encountered routinely.

The first is normal distribution (shown in Figure 13.13), which forms the foundation for much of the quality improvement work. Normal distributions are made up of **continuous data**.

Continuous data numbers that can be of any value or fraction of a value.

Given that description, you should not have used a normal distribution for your paper clip exercise because you did not capture fractional data. Chapter 14, *Variables Control Charts and Interpretation*, will focus on how to use control charts based on the normal distribution to distinguish common cause variability from special cause variability. It will show when to investigate your process for problems.

Another common data type that process technicians encounter is "counting data." Counting data are **discrete data** (variables data). For example, we charted the paper clips using only whole numbers; these numbers are discrete data.

Discrete data numbers that are not continuous (e.g., whole numbers). Also known as *variables data*.

Examples of counting data include spills, injuries, illnesses, and anything that is a count of events. Let us consider injuries for just a minute. You cannot have 1.5 injuries. You either had 1 injury or 2 injuries.

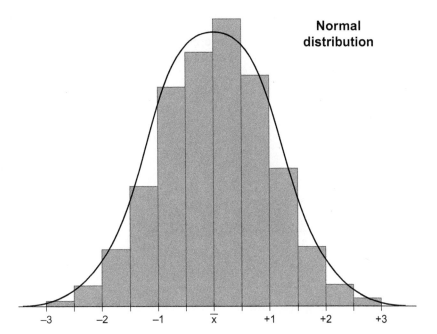

Figure 13.13 Normal distribution.

Given enough data, you might actually be able to apply a normal distribution to counting data. The appropriate distribution would be the **Poisson distribution** (shown in Figure 13.14).

Poisson distribution distribution of discrete data that consist of counting data. In this case, only whole numbers are present.

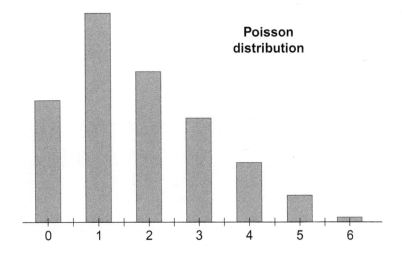

Figure 13.14 Poisson distribution.

When you look at a Poisson distribution, you will not see a smooth curve, because there are no data in between the whole numbers. In Chapter 15, *Attributes Control Charts and Interpretation*, you will learn how to use control charts based on a Poisson distribution.

The last commonly used distribution discussed at any length here is the **binomial distribution** (shown in Figure 13.15). Binomial literally means "consisting of two terms" (like yes/no). Like Poisson data, binomial data are discrete, not continuous.

What makes it look a little confusing is that the binomial distribution counts the number of failures, so the binomial distribution ends up looking a lot like the Poisson distribution. How to use control charts based on the binomial distribution is discussed further in Chapter 15, *Attributes Control Charts and Interpretation*.

Poisson and binomial distributed data are both referred to as attribute data. Rather than measuring some variable such as length or weight, these distributions attempt to measure a more generic attribute such as number of defects or the number of defective products produced.

Binomial distribution distribution of discrete data that are made up of data that have only two choices (e.g., pass/fail).

Figure 13.15 Binomial distribution.

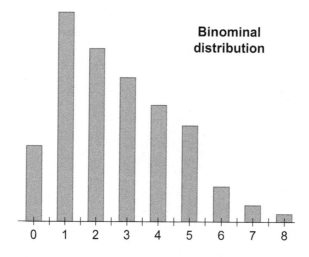

Binominal distribution

Summary

Consistency is a major part of the definition of quality. The opposite of consistency is variability. Because everything varies, variation is a natural part of all processes. By understanding the variability of a process and how that variation is distributed, you can use past data to predict future performance. You can also use the data to be alerted to the abnormalities in the process. Histograms are an excellent graphic tool to help in understanding the distribution of your data.

The distribution of data can be described using two key measures:

- Measures of central tendency—the most common being the mean or average

- Measures of spread—the most common being the standard deviation or sigma (σ)

The mean indicates where on the scale a process is centered, and the standard deviation indicates how much of the scale a process encompasses. For a normally distributed process, one can expect to find the following:

a. Approximately 68 percent of the data within 1 sigma of the mean

b. Approximately 95 percent of the data within 2 sigma of the mean

c. Approximately 99.7 percent of the data within 3 sigma of the mean

When the only variation seen in a process is the variation inherent in the process, this is known as common cause variation. A process is stable when it exhibits only common cause variation. A stable process is predictable.

When the variation in a process is over and above the natural, inherent variation of the process, this is known as special, or assignable, cause variation. A process is unstable when it exhibits special cause variation. An unstable process is unpredictable.

Normal distributions are made up of data that are continuous in nature; that is, any value or fraction of a value is allowed. These types of data are sometimes called variables data, as they represent measurements of many kinds of variables such as purity, weight, or viscosity.

Poisson distributions are made up of data that are discrete in nature; that is, only whole numbers are used. Poisson distributions are made up of counting data. Examples of counting data might include the counting of injuries, illnesses, or defects.

Binomial distributions are also made up of data that are discrete in nature, not continuous. Binomial distributions are made up of data that have only two options such as pass/fail. Examples of binominal data might include defective data on a production line.

Both Poisson and binomial data are sometimes called attribute data because they refer to an attribute of the product instead of a measure of a variable related to the product. Binomial and Poisson distributions look similar to each other because binomial data are captured by counting.

Checking Your Knowledge

1. Define the following key terms:
   a. Binomial distribution
   b. Common cause variation
   c. Histogram
   d. Mean
   e. Normal distribution
   f. Poisson distribution
   g. Range
   h. Special cause variation
   i. Stable process
   j. Standard deviation
   k. Unstable process

2. (*True or False*) Variability is the opposite of consistency.

3. Histograms are graphical tools used to:
   a. provide justification for capital projects.
   b. display the distribution of data in a tabular format.
   c. display the distribution of data in a bar chart format.
   d. provide management with information about shift performance.

4. Inherent variation:
   a. behaves in a predictable manner.
   b. is stable over time.
   c. is measurable.
   d. is random by nature.
   e. all of the above.

5. (*True or False*) Consistent operation is required if consistent product quality is desired.

6. Choose the options below that can help ensure consistent operation (Select all that apply):
   a. Standardized procedures
   b. Training
   c. Cheat sheets posted on the walls
   d. A degree in process technology
   e. Application of statistical tools to recognize special cause variation

7. (*True or False*) There are three key measures to understanding histograms.

8. Which is the most commonly used measure of central tendency?
   a. Mean
   b. Median
   c. Mode
   d. Middle

9. What is the most commonly used measure of spread?
   a. Range
   b. Standard deviation
   c. Variance
   d. Mode

10. Which of these Greek letters is used to denote the standard deviation?
    a. Σ
    b. μ
    c. σ
    d. Ω

11. Which of these Greek letters is used to denote the summation?

    a. Σ

    b. μ

    c. σ

    d. Ω

12. In a normal distribution, how much of the data is encompassed within 1 standard deviation of the mean?

    a. 33 percent

    b. 68 percent

    c. 95 percent

    d. 99.7 percent

13. In a normal distribution, how much of the data is encompassed within 2 standard deviations of the mean?

    a. 33 percent

    b. 68 percent

    c. 95 percent

    d. 99.7 percent

14. In a normal distribution, how much of the data is encompassed within 3 standard deviations of the mean?

    a. 33 percent

    b. 68 percent

    c. 95 percent

    d. 99.7 percent

15. A distribution of counting data is called:

    a. Boltzmann

    b. Normal

    c. Poisson

    d. Binomial

16. A distribution of pass/fail data is called:

    a. Normal

    b. Poisson

    c. Binomial

    d. Boltzmann

17. (*True or False*) A process that exhibits only common cause variation is said to be stable.

18. (*True or False*) A process that exhibits common cause and special cause variation is said to be stable.

NOTE: Answers to Checking Your Knowledge questions are in the Appendix.

Student Activities

1. Obtain 50 paper clips and complete the exercise outlined in the chapter, including the charting of the histogram. Be prepared to share your results in classroom discussion.

2. Create a colorful histogram using a large size bag of Skittles candy. Each Skittle represents a product defect. On a sheet of typing paper, separate Skittles into various categories based on color. The same color Skittles should be lined up in a column on top of each other. Afterwards draw a rectangular bar around each column of Skittles. On top of each bar, write the number of Skittles contained in the bar. Color categories should be placed on the x-axis. The y-axis should show the number of defects. To improve the process by reducing the number of defects, start with the bar with the highest number of defects.

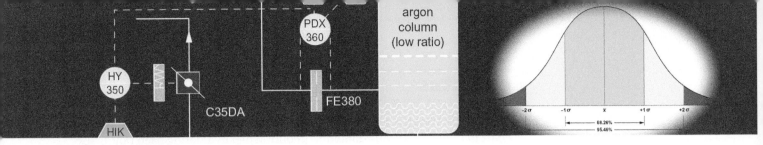

Chapter 14

Variables Control Charts and Interpretation

"The goods come back, but not the customer."

~ ROBERT W. PEACH (SEARS, ROEBUCK & CO.)

 ## Objectives

Upon completion of this chapter you will be able to:

14.1 Explain the purpose of variables control charts and how they relate to the distribution of data. (NAPTA Quality, Control Charts and Data Representation; Analysis and Interpretation 1*) p. 190

14.2 Given a data set, construct an x-bar and R control chart. (NAPTA Quality, Control Charts and Data Representation; Analysis and Interpretation 2, 3) p. 194

14.3 Given a data set, construct an individuals control chart. (NAPTA Quality, Control Charts and Data Representation; Analysis and Interpretation 2) p. 195

14.4 Explain the differences between individuals and x-bar and R control charts and when to use each one. (NAPTA Quality, Control Charts and Data Representation; Analysis and Interpretation 5) p. 197

14.5 Explain the purpose of other selected control charts:

zone chart

exponentially weighted moving average (EWMA) chart. (NAPTA Quality, Control Charts and Data Representation; Analysis and Interpretation 1) p. 199

*North American Process Technology Alliance (NAPTA) developed curriculum to ensure that Process Technology courses will produce knowledgeable graduates to become entry-level employees in process technology. Objectives from that curriculum are named here in abbreviated form. For example, "(NAPTA Quality, Control Charts and Data Representation; Analysis and Interpretation 1)" means that this chapter's objective 1 relates to objective 1 of the NAPTA curriculum about control charts and data representation.

Key Terms

Exponentially weighted moving average (EWMA) chart—a special type of control chart that takes previous data into account when evaluating each point; also capable of detecting small shifts in the mean, **p. 202**

Lower control limit (LCL)—bottom limit in quality control for data points below the control (average) line in a control chart, **p. 193**

Moving range (MR)—difference between two consecutive points, **p. 196**

Upper control limit (UCL)—the highest level of acceptable quality for a product or service, **p. 193**

Variables data—data from a distribution in which any number or fraction of a number is possible, **p. 190**

x-bar and R ($\overline{X}R$)—a control chart that displays both the mean value (X) as well as the range (R) to indicate changes in the mean value and dispersion over an established period of time, **p. 194**

14.1 Introduction

This chapter continues to focus on the statistical piece of the quality puzzle. It does this in relation to **variables data**.

Control charts are some of the most basic quality improvement tools. Walter Shewhart initially proposed their benefits in the 1920s and published the first work on the subject in 1939. Control charts are the main tools Dr. Deming employed in his work to help the Japanese rebuild their industrial infrastructure after World War II.

Control charts have been around a long time and have proven their worth over and over again. Other chapters in this section develop a thorough understanding of process variability and how that variability relates to the needs of customers. With that understanding as the foundation, we will start digging deeper into the various tools available to improve processes. Recall that Chapter 13, *Variance and Operating Consistency*, discussed different kinds of distributions used to describe different types of data. For continuous data, the normal distribution is used. The tools covered in this chapter are some of the many control charts, including the following:

- $\overline{x}$ bar and R charts
- Individuals charts
- Zone charts
- EWMA charts

As stated previously, there are many different tools in the quality toolbox, and each serves a different purpose. You need to select the right tool for the job in order to accomplish your goals as efficiently as possible. Selecting the right tool for the job requires an understanding of the purpose of the tools and a basic knowledge of when each should be applied. This chapter discusses the value of each type of control chart. You will also learn how to construct several of these charts so you can apply your knowledge to everyday situations.

Control Charts and Data Distribution
Purpose of Control Charts

Control charts serve a specific purpose in quality improvement efforts: They help distinguish between common cause and special cause variation. If you know the variation that you are seeing is inherent in the process, then you will accept it and keep moving forward. If you can see that the variation is not inherent in the process, then you know that something special has occurred. Your job will be to figure out what happened or assign a cause to the event.

Variables data data from a distribution in which any number or fraction of a number is possible.

Process technicians may assume that a special cause is a bad thing, but that may not be the case. If the special cause makes the process behave worse than normal, it is important to understand the cause to prevent its recurrence. If the special cause makes the process exhibit less variation than normal, then there is just as much need to investigate the issue. In this case, process technicians need to understand the cause so they can incorporate the change into their standard operating procedures to help their process operate at this improved level all the time.

Let us say that you are installing a new procedure or a new piece of equipment to make an improvement in your process. The control chart can tell you pretty quickly whether or not the improvement worked by showing whether the variation has changed or remained the same.

Relating Control Charts to Distributions of Data

Recall the normal distribution (shown in Figure 14.1). It is shaped like a bell, high in the center and tapering off to both sides. It is symmetrical, tapering quite evenly to the left and to the right.

Figure 14.1 Normal distribution.

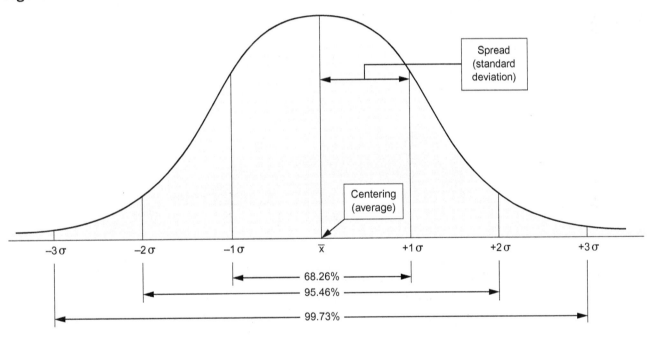

The variation to either side of the mean ($\bar{x}$) is distributed such that 68 percent of the numbers fall within 1 standard deviation (σ) of the mean, 95 percent fall within 2 standard deviations of the mean, and 99.7 percent fall within 3 standard deviations of the mean. For all intents and purposes, you would expect all of the process data to fall inside the 3 sigma limits. Only 3 times out of 1,000 would a point outside of the 3 sigma limits be normal.

The limitation in using this distribution is that a lot of data has to be collected before the curve can be drawn. It takes at least 20 to 30 data points to calculate the mean and standard deviation. If a process makes only one batch a day, it could take a month to collect enough data to describe the process with a distribution.

Distribution curves are useful, but histograms can provide more information about the process than a distribution curve can. Table 14.1 shows two lists of 100 numbers generated to be normally distributed with a mean of 30 and a standard deviation of 5. The same numbers are in each list. The set on the left is presented in random order and simulates the order in which these numbers would have been collected. The set on the right has been sorted from lowest to highest.

Table 14.1 List of 100 numbers in random and sorted order

Random										Sorted									
27	29	25	28	39	26	30	34	27	24	18	19	19	20	20	21	21	21	21	21
26	32	40	29	29	32	34	28	25	33	22	22	23	23	23	24	24	24	25	25
34	30	31	27	29	37	23	27	27	31	25	25	26	26	26	26	27	27	27	27
27	26	28	30	27	31	32	30	31	33	27	27	27	27	27	27	27	28	28	28
44	37	34	22	40	21	21	36	19	25	28	28	28	28	29	29	29	29	29	29
29	19	29	25	35	24	38	37	33	27	29	29	29	29	30	30	30	30	30	30
26	39	31	22	30	28	18	27	35	36	30	31	31	31	31	31	31	31	31	32
31	39	28	31	29	27	32	24	29	28	32	32	32	32	33	33	33	33	33	33
31	33	29	20	23	29	30	21	33	23	34	34	34	34	34	35	35	36	36	36
27	21	20	21	36	32	28	33	34	30	37	37	37	38	39	39	39	40	40	44

Both of these data sets would produce the exact same histogram (Figure 14.2) because the data are identical. The order in which the data were collected has no bearing on the distribution of the data. As you can see, the data are pretty much centered around 30 and appear to vary from roughly 15 to 45, which is what you would expect given a standard deviation of 5.

Figure 14.2 Histogram of data in Table 14.1.

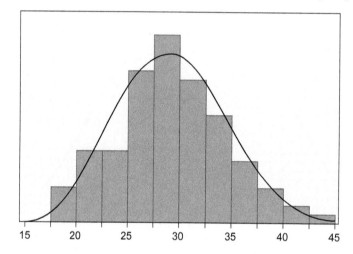

Notice that the histogram is not absolutely perfect. Even though these data were generated with computer software to force the mean and standard deviation, the random variability in the software functionality yielded a calculated mean of 29.3 and a calculated standard deviation of 5.1.

To add the element of *time* to our understanding of the distribution of these data, we will use a run chart or control chart. A run chart is just a sequential charting of the data. The random data from this example have been plotted (Figure 14.3). As you can see, the numbers

Figure 14.3 Run chart of data in Table 14.1.

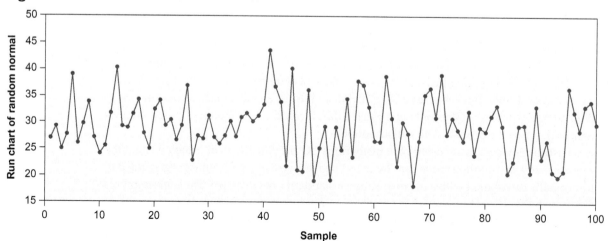

Sample

move up and down without anyone doing anything to the process. Nothing but common cause variation is seen here.

If you moved all these numbers to the left side of the graph, you would see that your linear graph of numbers can easily be related to the underlying normal distribution as shown in (Figure 14.4).

Figure 14.4 Run chart data showing a normal distribution.

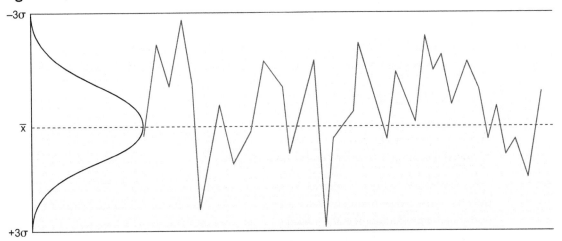

Because you know how to calculate the mean and standard deviation of these data, you do not have to present the histogram to show the distribution of the data. You can simply draw a line to represent the mean (average) of the data and draw lines to show the upper and lower limits of where you expect the data to be located (Figure 14.5). These upper and lower limits are called the **upper control limit (UCL)** and **lower control limit (LCL)**. They are traditionally set at the mean plus 3 standard deviation and the mean minus 3 standard deviation. By knowing this, you also know that 99.7 percent of the time your numbers should fall inside these limits.

Upper control limit (UCL)   the highest level of acceptable quality for a product or service.

Lower control limit (LCL)   bottom limit in quality control for data points below the control (average) line in a control chart.

Figure 14.5 Variable control chart data with upper and lower control limits.

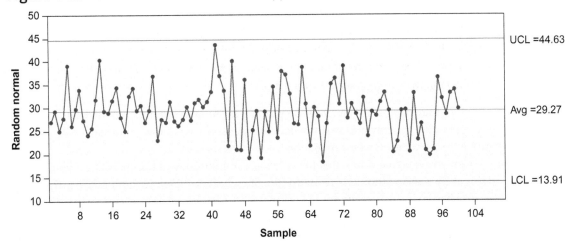

In Chapter 13, *Variance and Operating Consistency*, we laid out a simple example of how to calculate mean and standard deviation. For this current set of data, the calculations would look like this:

$$UCL = \bar{x} + 3\sigma = 29.3 + (3 \times 5.1) = 44.6$$
$$LCL = \bar{x} - 3\sigma = 29.3 - (3 \times 5.1) = 14.0$$

With a control chart in hand, there is no longer a need to wait to collect a minimum of 20 to 30 new data points to check on a process. Once control limits have been calculated, you can plot each new point as it is collected and check to see whether this point is "in control" or "out of control."

All of the control charts that will be discussed here are based on the same basic principles. First, we calculate control limits based on the underlying distribution of the data. Then we use these control limits to provide signals as to when the process is starting to behave differently than expected.

As mentioned before, different types of data have different underlying distributions and so they will require different kinds of control charts. Even data that come from the same underlying distribution may require different kinds of control charts. In the following sections we examine a couple of the most commonly used variables control charts.

14.2 Control Charts in Process Industries
X-bar and R ($\overline{X}R$) Control Charts

x-bar and R ($\overline{X}R$) a control chart that displays both the mean value (X) as well as the range (R) to indicate changes in the mean value and dispersion over an established period of time.

X-bar was described in Chapter 13, *Variance and Operating Consistency*. The **x-bar and R** control charts are used when data are grouped together. An x-bar and R control chart consists of two charts, one that depicts the average (x-bar) and one that depicts the range (R). In order to determine the control limits, the standard deviation must also be calculated.

Today handheld calculators or easily available software packages can calculate the standard deviation of a set of data. Process technicians are not expected to be able to calculate standard deviations by hand, but it is important to understand the idea behind the calculations. As review, here is an example of a set of five data points for individual measurements/observations (*x*): 2, 3, 9, 9, 7. As a reminder, here is the formula for the standard deviation presented in Chapter 13, *Variance and Operating Consistency*.

$$\sigma = \sqrt{\frac{\Sigma\,(x - \overline{x})^2}{n - 1}}$$

The standard deviation is calculated by doing the following (Figure 14.6):

1. Add up these five numbers to get 30, and then divide the total by 5 to get the mean (x-bar) = 6.

2. Subtract the mean (x-bar) from each individual measurement (*x*).

3. Square these values and then sum up (add) all the squares.

4. Divide this sum by the number of measurements you have (*n*) minus 1.

5. Take the square root of the entire value.

Using statistical software, you can put in a formula to calculate the standard deviation of a set of data quickly and easily. Many inexpensive calculators have statistical functions built in to calculate standard deviation.

Can you imagine doing this for a data set of hundreds of numbers? Can you imagine doing it by hand without a calculator like Shewhart and Deming and other early quality experts had to in their early days? Fortunately for them, they found a way to estimate the standard deviation without having to crank through all these numbers. They came up with a series of factors that would estimate the control limits based on the range (the difference between the high and low values) of data in a subset.

The earliest application of control charts was in what is sometimes called the widget industries, which make discrete things such as nuts, bolts, and tires that can be individually measured. Process technicians typically make drums or tanks of materials that have to be sampled and analyzed.

Count, n: 5

Sum, Σx: 30

Mean, μ: 6

Variance, σ^2: 8.8

$$\sigma = \sqrt{\frac{1}{n}\sum_{i=1}^{n}(x_i - \mu)^2}.$$

$$\sigma^2 = \frac{\Sigma(x_i - \mu)^2}{n}$$

$$= \frac{(2 - 6)^2 + \ldots + (7 - 6)^2}{5}$$

$$= \frac{44}{5}$$

$$= 8.8$$

$$\sigma = \sqrt{8.8}$$

$$= 2.9664793948383$$

Standard deviation $(\sigma) = 2.9664793948383$

Figure 14.6 Calculation of standard deviation for 2, 3 ,9, 9, 7.

The application of control charts when you make bins with hundreds of bolts in each is to sample a few bolts out of each bin and measure them. You capture the average measurement and also the range of the measurements. After you have captured 20 such samples, you can calculate the average of the ranges $(\overline{R})$.

Shewhart came up with a set of factors (A_2) that could estimate the control limits using this average range value. It is a set of factors instead of a single factor because the factor changes if you have two bolts in your sample versus three, four, or five bolts in a sample. The formulas he used for the upper and lower control limits looked like this:

$$UCL = \overline{x} + A_2\overline{R}$$
$$LCL = \overline{x} - A_2\overline{R}$$

Instead of adding and subtracting 3 standard deviations from the mean $(\overline{x})$ to get the control limits, you multiply the average of the ranges $(\overline{R})$ by some factor (A_2) based on the number of units in the sample. This becomes the estimate for the control limits. Tables of these factors (shown in Table 14.2) can easily be found in many books on control charting. With these numbers in hand, you can calculate control limits for your data.

Table 14.2 Factors for Estimating Control Limits

Number of Observations	Factor A_2
2	1.88
3	1.023
4	0.729
5	0.577

IN A NUTSHELL

$\overline{X}R$ charts are made from averages of rational subgroups of individual measurements. Use them when your data can be grouped rationally, for example, with SPC measurements where you can collect multiple measurements per batch.

14.3 Individuals Control Charts

Individuals control charts are plots of individual measurements instead of averages of subgroups. One perfect example of an appropriate application for an individuals control chart is your vehicle's gas mileage. You can collect data on your gas mileage and plot it on a control chart, but grouping tanks of gas together would not make any sense. What would you expect to see if you did this? Perhaps higher values would result when the entire tank of gas was expended on highway driving. Perhaps a string of low values would indicate that your vehicle is in need of a tune-up.

Oftentimes in the process industries there is no need to subgroup the data. For example, a chemical is made in a continuous process that fills up a tank every 12 hours. When that tank is full, a sample is taken for compositional analysis. When a passing analysis is obtained, the contents are transferred to a bulk storage tank.

If you are applying SQC techniques to the production facility, then you have a single set of analytical numbers to look at. You do not really have a grouping of numbers to average. For this reason, you may expect to find individuals control charts applied more frequently in the process industries than $\overline{X}R$ charts.

The basic methodology is the same:

1. Collect 20 to 30 data points so you can set the control limits.

2. Apply these control limits to new data that are collected and monitor the process for out-of-control signals.

3. The control limits can be calculated using the standard deviation, or they can be estimated using the range. In the case of individual measurements, there is not a subgroup of data to obtain a range, so the absolute value (always a positive number) of the **moving range (MR)** is used.

Moving range (MR) difference between two consecutive points.

In your process, you are making a chemical called dibutyl ether and are tracking the purity of this chemical. You make two tanks a day, so you have two purity analyses a day. If there is some rational reason to group the data from these two tanks together, then you might use an $\overline{X}R$ chart. However, if there is not a rational reason to group the data, then you can apply an individuals control chart instead. The beginning of your data set is shown in Table 14.3.

Table 14.3 Dibutyl Ether Data Set

Date	5/1	5/1	5/2	5/2	5/3	5/3	5/4	5/4
Batch No.	331	332	333	334	335	336	337	338
Result	98.2	98.8	99.4	99.5	98.7	99.4	98.3	99.8
MR		0.6	0.6	0.1	0.8	0.7	1.1	1.5

You have captured the purity values from each make tank and have calculated the moving range (MR) between the pairs. There cannot be a MR value for the first data point, and all of your MR values are positive because you use the absolute value of the difference. Some experts would suggest that plotting the MR as well as the $\bar{x}$ is the right way to go because a significant change in your ranges is also an indication of abnormality. Others would suggest that focusing on the variable of interest will tell you what you need to know and is much simpler to manage. For our use here, we will adopt the simpler option, but be aware that in industry you may see individuals control charts with a moving range section at the bottom, much like the $\overline{X}R$ charts showed.

The formulas for estimating the control limits for an individuals control chart based on the moving range are the same as $\overline{X}R$ charts. The A_2 factor for individuals control charts is always 2.66.

$$UCL = \bar{x} + 2.66 \times \overline{MR}$$
$$LCL = \bar{x} - 2.66 \times \overline{MR}$$

Following the preceding process, you would collect 20 to 30 data points and then calculate your average, $\bar{x}$, and your average moving range, $\overline{MR}$. The following 25 values are the data collected from this example:

99.0 98.5 99.4 99.5 99.8 99.8 99.2 99.7 99.0 99.6 99.1 99.4 99.7

98.9 99.5 99.0 98.6 99.5 99.2 99.4 99.6 99.3 99.5 98.9 99.6

Average these data to obtain an $\bar{x}$ of 99.3. Between each of the pairs of data, you calculate a moving range:

0.5 0.8 0.1 0.3 0.0 0.7 0.5 0.7 0.6 0.5 0.3 0.3

0.8 0.6 0.5 0.5 1.0 0.3 0.2 0.1 0.2 0.2 0.4 0.7

You average these moving ranges to obtain an $\overline{MR}$ of 0.5. Applying the control limit formulas, you calculate the UCL to be 100.6 and the LCL to be 98.0:

$$UCL = \bar{x} + 2.66 \times \overline{MR} = 99.3 + 2.66 \times 0.5 = 99.3 + 1.3 = 100.6$$

$$LCL = \bar{x} - 2.66 \times \overline{MR} = 99.3 - 2.66 \times 0.5 = 99.3 - 1.3 = 98.0$$

Notice that you have an UCL of 100.6. Should you plot an upper control of 100.6 when you are plotting the purity of a chemical? No, statistics is sometimes a funny business. The computers and calculators will provide numbers that are not always meaningful. In this case, something cannot be more than 100 percent pure. Just because you calculated a control limit above 100 percent does not make it rational. You would set your UCL at 100 because this is the practical limit for the measurement.

You can expect to see the same phenomenon when you are plotting impurities at low levels. In those cases, the LCL will sometimes be calculated as a number less than zero, but again, to have less than zero percent impurities is not possible, so you would set your lower control limit at zero. The result of your first control chart for this parameter (plotting the data and moving ranges above) is shown in Figure 14.7.

Look at this chart to see whether it looks "normal." The data seem to bounce closely around the average in a random pattern. Nothing appears abnormal, so you accept these control limits as normal for your process. Your next step would be to create a new control chart that already has these control limits drawn on it. Then you can continue your data collection and begin examining each new batch to monitor for process shifts.

14.4 X-Bar and R Control Charts *versus* Individuals Charts

$\overline{X}R$ charts are made from averages of rational subgroups of individual measurements. Use them when your data can be grouped rationally, for example, with statistical process control (SPC) measurements where you can collect multiple measurements per batch.

Individuals charts are made from individual measurements. Use them when your data have no rational subgrouping, for example, statistical quality control (SQC) applications where you have a single measurement per batch.

When deciding whether to use $\overline{X}R$ or individuals control charts, use the one that makes the most sense for your application. They are both valid tools and both serve a useful purpose. In fact, it is much more important to apply the tools you learn to improve your process than it is to worry about which tool may be the most technically correct tool for the job.

It is important to use the right tool, but sometimes there is more than one right answer. For example, you could use an open-ended wrench, a box wrench, or a socket to tighten a nut on a bolt. They are different tools, but they are all perfectly valid options. Just do not try to use a screwdriver. Results are what count. These quality technologies are just tools to help you get results.

IN A NUTSHELL

Individuals charts are made from individual measurements. Use them when your data have no rational subgrouping, for example, SQC applications where you have a single measurement per batch.

Figure 14.7 Sample individuals control chart.

ANY corp

Individuals control chart

Product: Dibutyl ether									Property: purity								Sample point: make tank							
Date	5/1	5/1	5/2	5/2	5/3	5/3	5/4	5/4	5/5	5/5	5/6	5/6	5/7	5/7	5/8	5/8	5/9	5/9	5/10	5/10	5/11	5/11	5/12	5/12	5/13
Batch No.	331	332	333	334	335	336	337	338	339	340	341	342	343	344	345	346	347	348	349	350	351	352	353	354	355
Result	99.0	98.5	99.4	99.5	99.8	99.8	99.2	99.7	99.0	99.6	99.1	99.4	99.7	98.9	99.5	99.0	98.6	99.5	99.2	99.4	99.6	99.3	99.5	98.9	99.6
MR		0.5	0.8	0.1	0.3	0.0	0.7	0.5	0.7	0.6	0.5	0.3	0.3	0.8	0.6	0.5	0.5	1.0	0.3	0.2	0.1	0.2	0.2	0.5	0.7
Observation	1	2	3	4	5	6	7	8	9	10	11	12	13	14	15	16	17	18	19	20	21	22	23	24	25

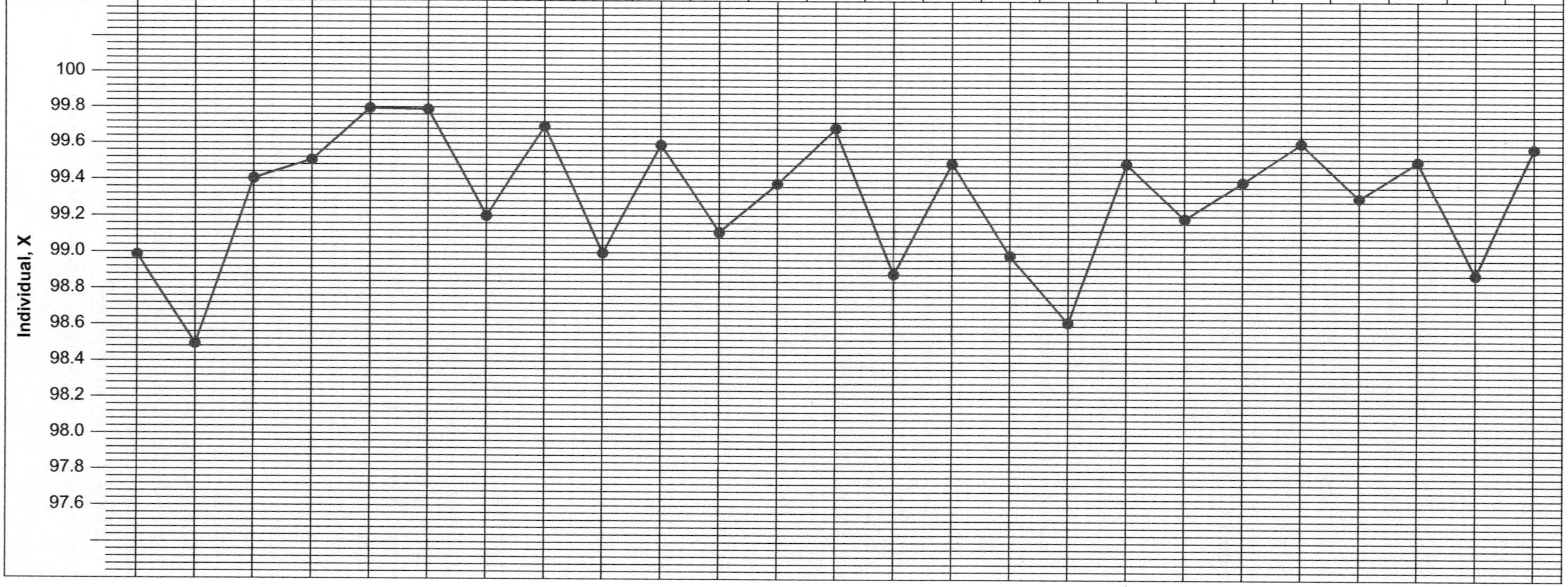

14.5 Other Selected Control Charts

Zone Charts

A technique that has been developed relatively recently is that of the zone chart (Figure 14.8). It is basically an individuals control chart that has been made even easier to plot and manage.

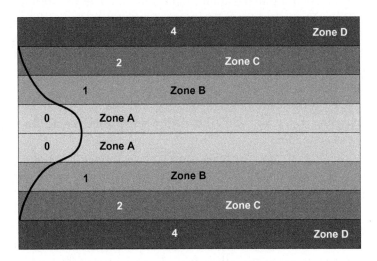

Figure 14.8 Sample zone chart.

Think back to the discussion about variability concepts: 68 percent of your data fall within 1σ of your mean, 95 percent fall within 2σ, and 99.7 percent fall within 3σ. Zone charts are designed on the premise that you can then predict how frequently data should fall within each of these sigma zones. You can set a numerical value to data that fall within each zone.

There are many scoring systems in use, depending on how reactive the user wants the control chart to be. The most commonly used system is a score of zero for data that fall within 1σ, a score of 1 for data that fall within 2σ, a score of 2 for data that fall within 3σ, and a score of 4 for any data that fall outside the 3σ control limits. A zone chart template might look like the example shown in Figure 14.9.

Figure 14.9 Example of a zone chart template.

ANY corp   Zone chart

Product:											Property:							Sample point:					
Date																								
Batch No.																								
Results																								
Score																								
Observation	1	2	3	4	5	6	7	8	9	10	11	12	13	14	15	16	17	18	19	20	21	22	23	24	25

Out of control		4
Zone C		2
Zone B		1
Zone A		0
Zone A		0
Zone B		1
Zone C		2
Out of control		4

Setting up the control chart is the same as setting up an $\overline{X}R$ or individuals control chart. First you have to collect 20 to 30 data points. Then you calculate the zone limits or estimate them based on the range or moving range. With this information in hand, you can construct the control chart for your future monitoring. Here is where the zone chart gets easier to use.

When plotting data on a zone chart, you do not have to worry about the exact scale—just plot the point inside the appropriate zone based on the standard deviation calculation or estimate. Every time you collect a new data point, ask yourself the question, "Did my plotted line just cross the mean?" If so, simply record the score for the new data point. If not, add the score of the new point to the existing score and record the sum. Whenever your score exceeds 4, your control chart is giving you a signal that something is wrong and you should investigate to determine the special cause.

Let us construct an example using the data from the individuals control chart on the preceding pages. You will estimate the sigma zones using the moving range information already provided.

Recall that your $\overline{x}$ was 99.3 and your $\overline{MR}$ was 0.5. You used the following formula to set your upper control limit at 100.6: $UCL = \overline{x} + 2.66 \times \overline{MR}$. You then used the following formula to set your lower control limit at 98.0: $LCL = \overline{x} - 2.66 \times \overline{MR}$.

Because you know that these control limits represent the $+3\sigma$ and -3σ limits, you can easily calculate the single standard deviation by taking the difference between the UCL and the $\overline{x}$ and dividing by 3. You can then add this single standard deviation to the mean and subtract it from the mean three times to set the rest of the zone limits.

For this case, you already calculated that the UCL was 100.6. Subtracting your mean (99.3) from this and dividing by 3 tells you that a single standard deviation is estimated to be $(100.6 - 99.3)/3 = 1.3/3 = 0.433$. You fill in the zone chart template with the $\overline{x}$ of 99.3 and with your zone limits of $99.3 \pm 1 \times 0.433$, $99.3 \pm 2 \times 0.433$, and $99.3 \pm 3 \times 0.433$. Try this on your own and see whether you get zone limits like the ones shown in Figure 14.10.

Figure 14.10 Zone chart limits of dibutyl ether.

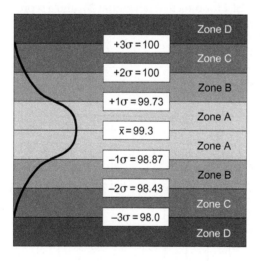

If these limits had already been put in place, and then you collected the same data that were previously shown in your individuals control chart example, then your zone chart would have looked like the one shown in Figure 14.11.

Notice that because your process is running so close to the practical limit of 100 percent, you rarely achieved a high score. You keep adding the new score to the old score until your line crosses the mean, then you start over with your score.

Zone charts exist to make setting up, plotting, monitoring, and interpretation of control charts as simple as possible. They do this by having you plot your data within sigma zones instead of on an exact numerical scale and by replacing the more complex interpretation rules (which will be discussed in Chapter 15, *Attributes Control Charts and Interpretation*) with a simpler scoring mechanism.

Variables Control Charts and Interpretation 201

Figure 14.11 Zone chart of dibutyl ether.

ANY corp

Zone chart

Product: Dibutyl ether

Property: purity

Sample point: make tank

Date	5/1	5/1	5/2	5/2	5/3	5/3	5/4	5/4	5/5	5/5	5/6	5/6	5/7	5/7	5/8	5/8	5/9	5/9	5/10	5/10	5/11	5/11	5/12	5/12	5/13
Batch No.	331	332	333	334	335	336	337	338	339	340	341	342	343	344	345	346	347	348	349	350	351	352	353	354	355
Results	99.0	98.5	99.4	99.5	99.8	99.8	99.2	99.7	99.0	99.6	99.1	99.4	99.7	99.9	99.5	99.0	99.6	99.5	99.2	99.4	99.6	99.8	99.5	99.9	99.6
Score	0	1	0	0	1	2	0	0	0	0	0	0	0	0	0	0	1	0	0	0	0	0	0	0	0
Observation	1	2	3	4	5	6	7	8	9	10	11	12	13	14	15	16	17	18	19	20	21	22	23	24	25

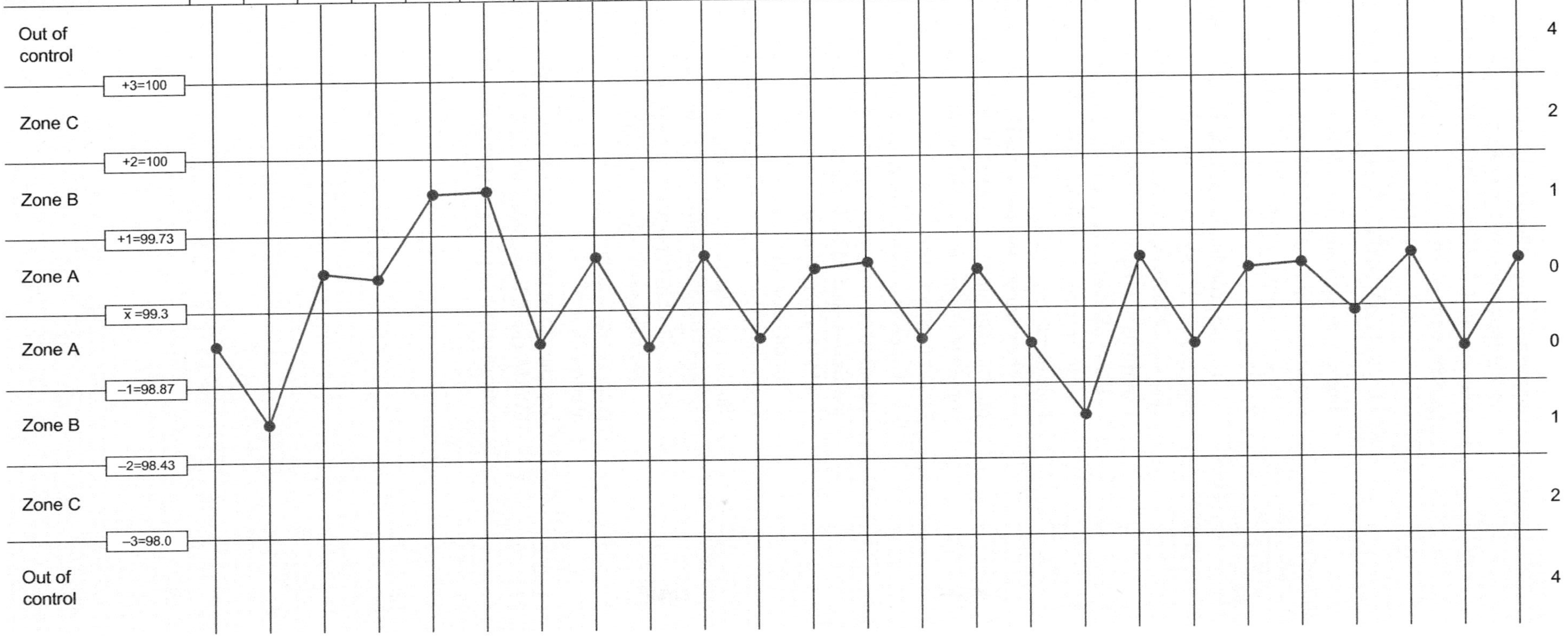

Out of control — 4

Zone C — 2 — +3=100

Zone B — 1 — +2=100

Zone A — 0 — +1=99.73

— 0 — $\bar{x}$=99.3

Zone A — 0 — −1=98.87

Zone B — 1 — −2=98.43

Zone C — 2 — −3=98.0

Out of control — 4

The types of charts explored so far are expected to be constructed with data that are completely independent of each other and for a process that is stable over time. This is not always the case, so you may see a couple of specialty control charts in the workplace. Because they are much more difficult to set up and operate, the only goal in this book is to explain in general terms what they are so you will recognize them when you see them again.

Exponentially Weighted Moving Average (EWMA) Charts

Traditional Shewhart control charts use the last data point plotted for making decisions. An **exponentially weighted moving average (EWMA) chart** is another way to take previous data into account when evaluating each point and making "in-control" or "out-of-control" decisions.

In many continuous processes, raw materials are flowing into a reactor at the same time final product is flowing out the other side. You will learn more about residence time in other classes, but for now, just consider that if you are putting 10 gallons per minute into a reactor that holds 3,600 gallons, then it will take 6 hours to fill the reactor. Once the reactor is full, you do not change out the entire volume of the reactor every 6 hours because some of what you are putting in is mixing with what was already there and not flowing straight through (turbulent flow versus plug flow).

The engineering rule of thumb is that it takes $2\frac{1}{2}$ residence times to completely change out the volume of the vessel. For this example, that means you need 15 hours to make a complete change in the reactor contents. This particular reactor is sampled every 6 hours for compositional analysis.

If a process technician is not careful, he or she could make a change to the process conditions, then take a sample 4 hours later, obtain a result that is not consistent with the new conditions, and so be fooled into making another change in the process conditions. This would be a drastic mistake because the change has not been given time to take effect yet.

In this case, the data that you are collecting every 6 hours are not totally independent of each other. The sample taken 6 hours after a process change will show only a little effect of the change. The 12-hour sample will show most of the change. The 18-hour sample will be completely representative of the change being implemented. Each data point is related to the last couple of data points, but more heavily related to the most recent point.

For cases such as these, the best choice is an EWMA chart, where you plot the individual points as well as the exponentially weighted moving average, but you set your control limits around the EWMA and use those numbers to make your decisions. EWMA charts are also designed to be sensitive to small changes in the mean. Figure 14.12 shows an example of an EWMA chart. Remember, for an EWMA chart you are not concerned when an individual point is out of control, but only when the EWMA is out of control.

Exponentially weighted moving average (EWMA) chart a special type of control chart that takes previous data into account when evaluating each point; also capable of detecting small shifts in the mean.

Figure 14.12 Example of an exponentially weighted moving average (EWMA) chart.

ANY corp

Zone chart

Sample point: make tank

Property: purity

Product: Dibutyl ether																								
Date	5/1	5/1	5/2	5/2	5/3	5/3	5/4	5/4	5/5	5/5	5/6	5/6	5/7	5/7	5/8	5/8	5/9	5/9	5/10	5/10	5/11	5/11	5/12	5/12	5/13
Sample number	331	332	333	334	335	336	337	338	339	340	341	342	343	344	345	346	347	348	349	350	351	352	353	354	355
Results	99.0	98.5	99.4	99.5	99.8	99.8	99.2	99.7	99.0	99.6	99.1	99.4	99.7	99.9	99.5	99.0	98.6	99.5	99.2	99.4	99.6	99.8	99.5	99.9	99.6
EWMA			99.1	99.4	99.7	99.8	99.4	99.6	99.2	99.4	99.2	99.3	99.8	99.5	99.3	99.2	98.8	99.2	99.2	99.4	99.5	99.4	99.4	99.1	99.4
Observation	1	2	3	4	5	6	7	8	9	10	11	12	13	14	15	16	17	18	19	20	21	22	23	24	25

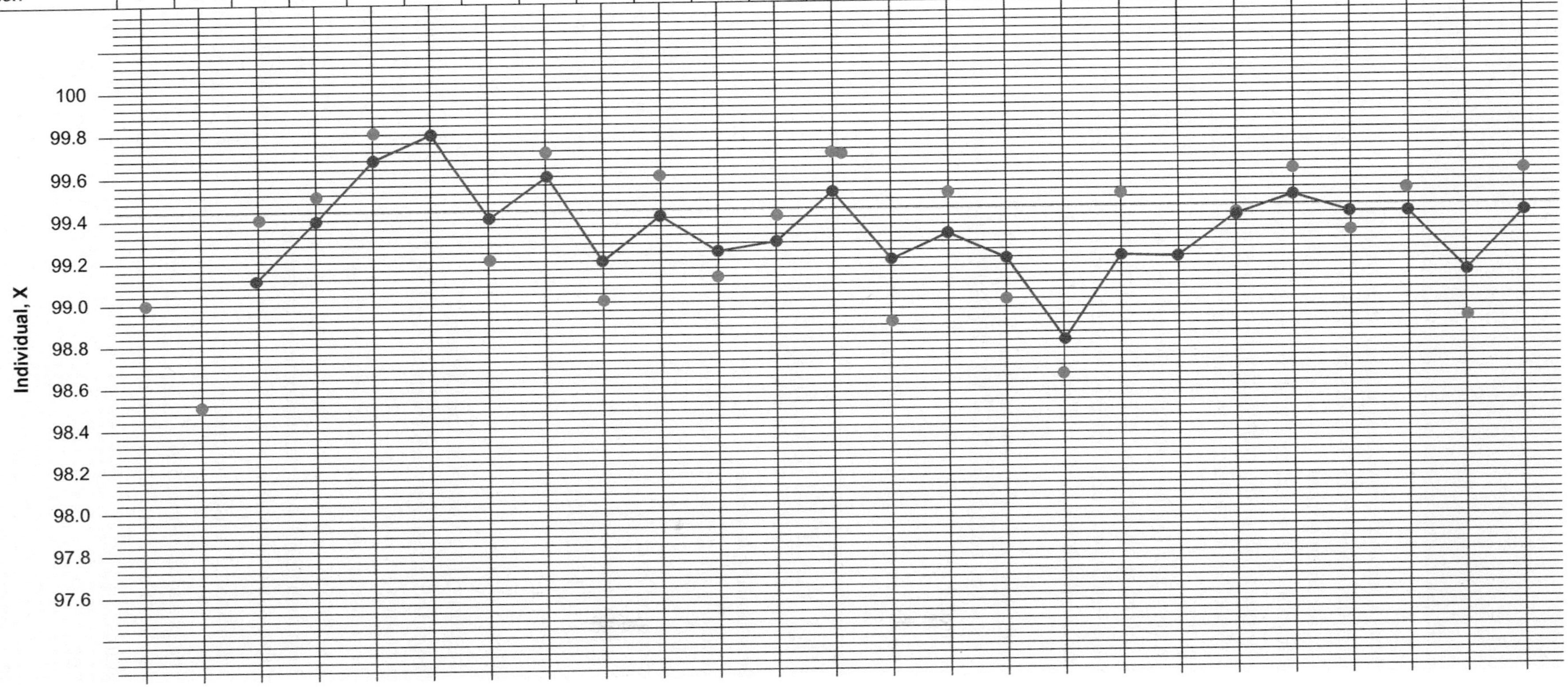

Summary

Control charts are some of the most basic quality improvement tools. Some were developed many decades ago, whereas others are much more contemporary. Control charts are a mechanism to look at your data in real time and to use the data to monitor your process. All of the control charts discussed in this chapter are based on variables or continuous data and the normal distribution.

If your process data can be rationally divided into subgroups, then you can apply an x-bar and R ($\overline{X}R$) control chart. If your data are structured such that they would better be examined individually, then you should apply an individuals control chart.

If you would like to implement the simplest possible control chart, then you could substitute a zone chart for an individuals control chart.

If you have a specific need to detect small shifts in the process mean, you could apply EWMA charts. If your data are not independent of each other, then by all means you should be applying the EWMA chart.

As has been stated previously, there are many different tools in the quality toolbox and each serves a different purpose. You need to select the right tool for the job to accomplish your goals as efficiently as possible. Selecting the right tool for the job requires an understanding of the purpose of the tools and a basic knowledge of when each should be applied.

Checking Your Knowledge

1. Define the following key terms:
   a. Exponentially weighted moving average (EWMA) chart
   b. Lower control limit (LCL)
   c. Moving range (MR)
   d. Upper control limit (UCL)
   e. Variables data
   f. x-bar and R ($\overline{X}R$)

2. What is the purpose of variables control charts?
   a. They help distinguish between common cause and special cause variation.
   b. They help distinguish between uncommon cause and special cause variation.
   c. They help distinguish between uncommon cause and assignable cause variation.

3. (*True or False*) Special causes are always bad.

4. Control charts add the element of _____ to your understanding of the distribution of data.
   a. histograms
   b. variation
   c. time
   d. money

5. What is the difference between individuals control charts and $\overline{X}R$ charts?
   a. There is no difference—they are interchangeable.
   b. Individuals control charts use individual measurements, whereas $\overline{X}R$ charts use averages of measurements divided into subgroups.
   c. Individuals control charts use averaged measurements.
   d. Individuals control charts cannot be used by teams.

6. You should use $\overline{X}R$ charts:
   a. when there are more than 100 data points.
   b. when your data are grouped into rational subgroups.
   c. when your data cannot be grouped into rational subgroups.
   d. as the default when you do not know what else to do.

7. You should use individuals control charts:
   a. when you suspect individual trends are occurring.
   b. when your data are grouped into rational subgroups.
   c. when your data cannot be grouped into rational subgroups.
   d. as the default when you do not know what else to do.

8. What is the purpose of zone charts?
   a. They make setting up, plotting, monitoring, and interpretation of control charts as simple as possible.
   b. They are the technically correct charting technique when you are in the quality zone.
   c. They can detect small shifts in the mean.
   d. They take all previous data into account when determining how good your process is.

9. A control chart that takes previous data into account but gives more weight to the most recent data is the _____ chart.
   a. EWMA
   b. Time
   c. FMEA
   d. MAIC

10. Which is the best control chart to use in the processing industries?
   a. $\overline{X}R$
   b. Zone
   c. EWMA
   d. Individuals
   e. Whichever one best fits the particular need

11. EWMA, zone, individuals, and $\overline{X}R$ charts are all based on the _____ distribution.
   a. Poisson
   b. binomial
   c. normal
   d. rectangular

12. (*True or False*) You need to understand the purpose of the various quality tools available to choose the right tool for the right job at the right time.

NOTE: Answers to Checking Your Knowledge questions are in the Appendix

Student Activities

1. You are operating a process that makes approximately 150 gallons of dibutyl ether. Due to the changes in the viscosity of this product as it is drummed, you must sample each drum as it is filled. Because each drum within a batch is logically linked to the other drums, you decide to implement an $\overline{X}R$ chart for this activity. Given the initial data below, calculate the $\bar{x}$ and range for each batch, the mean and $\overline{R}$ for the entire data set, and the control limits you would expect to set for this activity. Using the template your instructor will supply, plot your data with the control limits. What can you conclude from your initial chart?

Observation	Drum 1	Drum 2	Drum 3
1	55.8	56.8	55.8
2	54.3	53.3	53.4
3	52.7	52.4	54.8
4	53.7	54.9	59.3
5	55.8	54.7	55.0
6	57.8	57.2	55.5
7	54.2	53.5	54.8

Observation	Drum 1	Drum 2	Drum 3
8	57.0	57.6	57.8
9	57.4	57.3	59.5
10	54.1	55.6	56.6
11	56.1	55.4	57.6
12	57.6	59.1	59.9
13	48.3	62.0	58.1
14	56.1	56.6	57.2
15	54.0	55.6	56.5
16	57.5	56.8	54.1
17	54.1	54.6	58.2
18	57.4	56.4	53.9
19	57.3	56.5	55.4
20	53.2	54.2	55.3
21	55.2	53.3	52.3
22	55.4	56.9	58.1
23	55.8	56.6	58.2
24	53.7	55.4	52.1
25	55.5	55.3	57.3

2. In the process of making dibutyl ether, you also check the modulus (measurement of product quality) of the chemical in the reactor. You get only one measurement per batch, so you implement an individuals control chart. Your initial set of data is provided below. Calculate the $\bar{x}$ and moving range for these data, the mean and $\overline{MR}$ for the entire data set, and the control limits that you would expect to set for this activity. Using the template your instructor will supply, plot your data with the control limits. What can you conclude from your initial chart?

Observation	Modulus
1	152.7
2	152.4
3	152.6
4	151.6
5	152.8
6	152.6
7	152.9
8	151.5
9	152.6
10	154.1
11	153.7
12	151.7
13	152.2
14	153.9
15	151.6
16	152.0
17	151.0
18	151.5
19	153.1
20	152.1
21	151.1
22	153.7
23	152.8
24	152.6
25	152.7

3. After completing your individuals control chart above, you implement this chart and use it to monitor your process for the next 2 months. At the end of that time, you rerun all the calculations. You determine that the process mean is 153 and the single standard deviation is 0.8. Using these numbers, calculate the zone limits for your process, and plot the following data on the zone chart template your instructor will supply. What can you learn from this chart? How is your process doing?

Observation	Modulus
1	152.7
2	152.4
3	152.6
4	151.6
5	152.8
6	152.6
7	152.9
8	151.5
9	152.6
10	154.1
11	153.7
12	151.7
13	152.2
14	153.9
15	151.6
16	152.0
17	151.0
18	151.5
19	153.1
20	152.1
21	151.1
22	153.7
23	154.8
24	155.2
25	154.7

Chapter 15

Attributes Control Charts and Interpretation

"The goal is to turn data into information, and information into insight."

~ Carly Fiorina (Hewlett-Packard)

Objectives

Upon completion of this chapter you will be able to:

15.1 Explain the purpose of attributes control charts and how they relate to their specific distributions of data. (NAPTA Quality, Basics of SPC 5*) p. 208

15.2 Explain what characterizes Poisson C and U control charts, as well as when and how to construct each type. (NAPTA Quality, Control Charts and Data Representation; Analysis and Interpretation 4) p. 210

15.3 Explain what characterizes binomial NP and P control charts, as well as when and how to construct each type. (NAPTA Quality, Control Charts and Data Representation; Analysis and Interpretation 4) p. 214

15.4 Apply the three main rules for interpreting control charts to determine the presence of special causes. (NAPTA Quality, Control Charts and Data Representation 1, 6) p. 219

*North American Process Technology Alliance (NAPTA) developed curriculum to ensure that Process Technology courses will produce knowledgeable graduates to become entry-level employees in process technology. Objectives from that curriculum are named here in abbreviated form. For example, "(NAPTA Quality, Basics of SPC 5)" means that this chapter's objective 1 relates to objective 5 of the NAPTA curriculum about statistical process control.

Key Terms

Attributes data—data from a discrete distribution in which only whole numbers (counts) are possible, **p. 208**

C chart—control chart for counts of defects (Poisson distribution) with a constant sample size, **p. 210**

NP chart—control chart for counts of defectives (binomial distribution) with a constant sample size, **p. 214**

Out of control—points outside the three sigma (σ) control limits, **p. 220**

P chart—control chart for counts of defectives (binomial distribution) with a variable sample size, **p. 214**

Shifts—runs of eight points above or below the average, **p. 220**

Trends—direction in a set of statistical data at a particular time; runs of eight points going up or going down, **p. 221**

U chart—control chart for counts of defects (Poisson distribution) with a variable sample size, **p. 210**

15.1 Introduction

The majority of the discussion thus far has been about data based on the normal distribution as well as the proper application of quality tools to those data, such as control charts. This type of data is called variables data. Variables data are data from a distribution in which any number or fraction of a number is possible.

In your role as a process technician, you will frequently encounter other types of data. For example, every process plant tracks safety statistics such as injuries, illnesses, and spills. It also tracks quality data such as rejected batches and customer complaints.

These types of non–normally distributed data are collectively referred to as **attributes data**. Attributes data are typically counted as whole numbers. However, you will see later in this chapter that you may count whole numbers and then report fractions.

The purpose of attributes control charts is the same as the purpose of variables control charts—to distinguish between common cause and special cause variation. The only difference is that because the data have a different underlying distribution, the control limits have to be calculated differently. This chapter explores control charts that are based on various attributes distributions. It also defines some basic rules for interpreting both variables control charts and attributes control charts.

Attributes data data from a discrete distribution in which only whole numbers (counts) are possible.

Attributes Distributions

When it comes to health, safety, environmental, and quality data, process technicians are typically counting items or events. They count injuries, customer complaints, spills, and product rejects. The counted data generally fall into distributions called the Poisson distribution or the binomial distribution. The Poisson distribution is the underlying distribution for the counting of defects. The binomial distribution is the underlying distribution for the counting of defectives. Until you have heard them a few times and seen them in practice, these descriptions sound much too similar to distinguish, but they are different. Let us look at each one separately to make the distinction clearer.

Defects *versus* Defectives

You will remember from Chapter 13, *Variance and Operating Consistency*, that *Poisson distribution* is a discrete distribution, made up of whole numbers only (Figure 15.1). It represents the counts of *defects*.

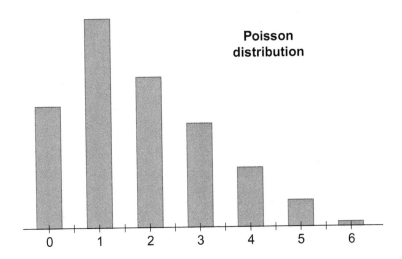

Poisson distribution

If you define your employee population as the target of your study, then every injury or illness is a defect. If you define a single drum of chemical as the target of your study, then every dent, paint chip, or crooked label could be defined as a defect. In each of these cases, what matters is that you define what you are studying so that you can apply the appropriate improvement tool.

In each of these cases, you can have multiple defects counted and still have an acceptable working process. You can experience one or two injuries per year and still operate your plants. You might receive a couple of customer complaints about dented drums or scratched labels, but you can continue to sell product to these customers. These counts indicate your process is not perfect and perhaps provide information about how to improve your process, but they do not invalidate the entire process.

A *binomial distribution* is a discrete distribution, made up of whole numbers only. It represents the counts of *defectives*.

If you define each person in your employee population as the target of your study, then any person who is unable to fulfill his or her job duties due to an injury is defective. If you define a single drum of chemical as the target of your study, then you must examine it to determine whether it is defective. If you can ship it, then it may contain defects but it is not defective. If you cannot ship the drum, then you must define it as defective. If you define your process as delivering quality goods to the customer, then any batch of product that cannot be sold is defined as defective. You reject this product and do not ship it to the customer.

In each of these cases, what matters is that you define what it is you are studying so you can apply the appropriate improvement tool. In the case of the Poisson distribution, the object of your study was not perfect, and you counted the defects. In the case of the binomial distribution (shown in Figure 15.2), the object of your study was product unacceptable for shipment to the customer, and you counted it as defective.

An easy way to remember the options for a binomial distribution is that there are only two options: pass (accept) or fail (reject). *Bi-* means two. A binomial distribution has two answers, just as a bicycle has two wheels. It is clear from Figure 15.2 that the level of defectives was high.

How often do process technicians need to collect their counting data? The answer depends on what they are trying to accomplish. For example, if the challenge is to improve the annual performance of a manufacturing facility that has a 20-year history of quality problems, they might start with an analysis of the annual customer complaint data.

If the goal is to assess a newer plant that has been running for only 5 years, it probably makes sense to start with a quarterly or monthly data analysis. This is because there are not enough annual data to be of value. Counting customer complaints per month or per quarter

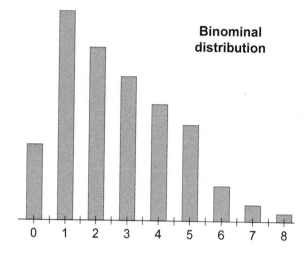

Figure 15.2 Example of a binomial distribution showing the number of defectives that were unable to be sold.

is a fairly common practice. Counting customer complaints per day or per week would result in too little information included in each count.

When collecting data, sometimes the sample size is constant and sometimes the sample size is variable. For example, let us count the number of customer complaints per month. This sounds like a constant sample size; but if the company made 50 shipments last month and only 25 this month, a count of two complaints in each month carries a different meaning. Two complaints out of 25 shipments is a much worse performance than two complaints out of 50 shipments. For each of the two distributions just discussed, there are two different control charts: one for a variable sample size and one for a constant sample size. Let us take a look at them.

15.2 Control Charts for Poisson Distributed Data

If the data are Poisson distributed, the next step is to examine the source of the data to determine whether the sample size is constant or whether it varies. If the sample size is constant, then the appropriate control chart is a **C chart**. If the sample size varies, then the appropriate control chart is a **U chart**.

C chart   control chart for counts of defects (Poisson distribution) with a constant sample size.

U chart   control chart for counts of defects (Poisson distribution) with a variable sample size.

The overall methodology for attributes control charting is the same as for variables charting.

1. Collect a reasonable amount of data to estimate the control limits.
2. Plot the data versus time to determine what you might see.
3. Impose the control limits on the plotted data.
4. Evaluate the data to determine whether there are special causes present.
5. When satisfied with the initial data collection, set up blank charts using these control limits to evaluate future data as they are collected.

C Charts

Because a C chart is much simpler than a U chart, we will start there. The standard deviation for a Poisson distribution is the square root of the mean, or average. The formula looks like this:

$$\text{Sigma}_C = \sqrt{\overline{C}}$$

The control limits for a C chart are the same as for variables charts. The control limits are set at 3 standard deviations above and below the average ($\overline{C}$). However, a standard deviation calculation for a Poisson distribution is different from a standard deviation calculation for a normal distribution. The combination of the standard deviation formula with the 3 standard deviation control limits results in the following formulas for the control limits of a C chart:

$$UCL = \overline{C} + 3\sqrt{\overline{C}}$$
$$LCL = \overline{C} - 3\sqrt{\overline{C}}$$

Let us suppose a company produces isopropyl alcohol, and upper level managers are wondering how the business is doing commercially. The monthly data for the last couple of years would be gathered in order to have at least 20 data points.

Your effectiveness each month is evidenced by the number of customer complaints for that month *versus* the number of sales. So, a count of the number of customer complaints received each month would be done. Table 15.1 shows the result.

Table 15.1 Customer Complaints over a Two-Year Period

Month/Year	Number of Defects	Month/Year	Number of Defects
01/20	15	01/21	3
02/20	8	02/21	6
03/20	10	03/21	4
04/20	7	04/21	4
05/20	4	05/21	9
06/20	9	06/21	2
07/20	1	07/21	2
08/20	9	08/21	2
09/20	7	09/21	5
10/20	3	10/21	2
11/20	5	11/21	4
12/20	4	12/21	4

With these data in hand, you pull up your attributes chart template and fill in the numbers. Then, you add up all these complaints (the sum is 129) and divide by the number of counts that you have (24) to get the average, or $\overline{C}$, which you calculate to be 5.4. The square root of $\overline{C}$ is the standard deviation for your data. This comes out to 2.3. Your control limits are set at 3 standard deviations (3 × 2.3 = 6.9) above the average and 3 standard deviations below the average. The following are the control limit formulas with the numbers included.

$$UCL = \overline{C} + 3\sqrt{\overline{C}} = 5.4 + (3 \times 2.3) = 5.4 + 6.9 = 12.3$$
$$LCL = \overline{C} - 3\sqrt{\overline{C}} = 5.4 - 3(3 \times 2.3) = 5.4 - 6.9 = -1.5 = 0$$

(**Note about the LCL**: A control limit with a negative number does not make sense. You cannot have fewer complaints than zero), so you would use ZERO as your lower control limit.)

You can now plot your monthly counts of customer complaints and add lines for the average and the upper control limit. The result should look like the chart shown in Figure 15.3.

What conclusions can be drawn from this chart? Are things getting better or are they getting worse over this time frame? What if that first data point were not present? Would you draw the same conclusion, or is that one piece of data skewing your perception of how

Figure 15.3 Control chart of the number of customer complaints (see vertical label on left).

Month/Yr	01-20	02-20	03-20	04-20	05-20	06-20	07-20	08-20	09-20	10-20	11-20	12-20	01-21	02-21	03-21	04-21	05-21	06-21	07-21	08-21	09-21	10-21	11-21	12-21	01-22
No. defects	15	8	10	7	4	9	1	9	7	3	5	4	3	6	4	4	9	2	2	2	5	2	4	4	5
Sample size																									
No./sample																									
UCL	12.3	12.3	12.3	12.3	12.3	12.3	12.3	12.3	12.3	12.3	12.3	12.3	12.3	12.3	12.3	12.3	12.3	12.3	12.3	12.3	12.3	12.3	12.3	12.3	12.3
LCL	0	0	0	0	0	0	0	0	0	0	0	0	0	0	0	0	0	0	0	0	0	0	0	0	0
Observation	1	2	3	4	5	6	7	8	9	10	11	12	13	14	15	16	17	18	19	20	21	22	23	24	25

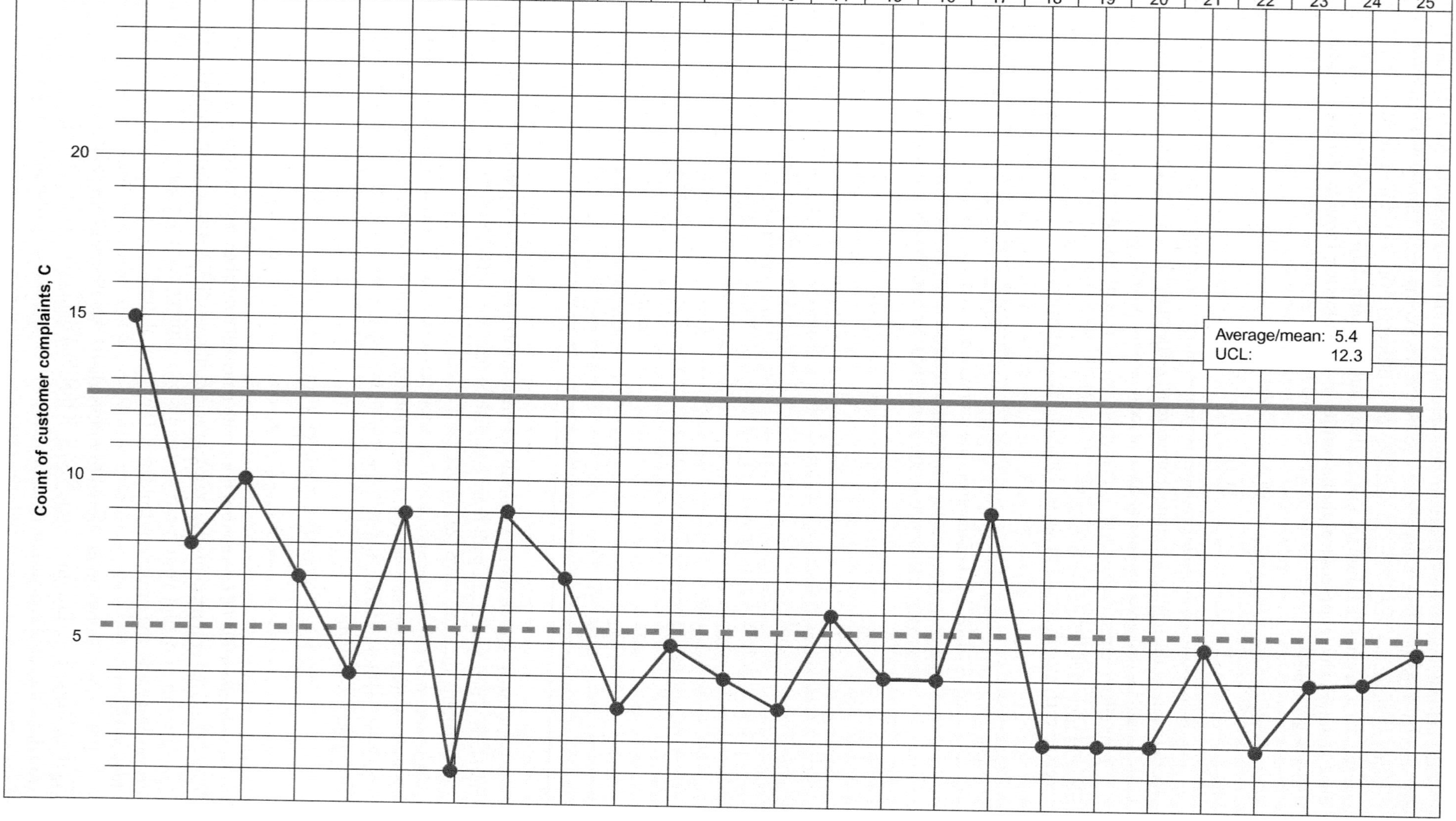

Business unit: Isopropyl alcohol (worldwide ops)

Chart: customer complaints

Description: minor & major

ANY corp

Attributes chart © U P NP

Average/mean: 5.4
UCL: 12.3

Count of customer complaints, C

this work process is performing? What about the last seven data points on this chart? Do these points suggest any trend in the data? Soon we will discuss tools to help interpret what the control chart is indicating.

You might notice in Figure 15.3 that there is a place to label the data. In this case, row 1 indicates the month and year, and row 2 is the place to collect the count of defects. The upper and lower control limits are also shown. A couple of other sets of information are not provided These would be used to collect the sample size and the number of defects per sample. These fields are needed only when your sample size is not constant.

U Charts

If the count of defects has a *variable* sample size, then a U chart is the proper chart to use. Two factors make a U chart more difficult to deal with than a C chart: (1) the control limits vary based on sample size, and (2) the sample size varies with every data point collected. For the sake of completeness, we are providing the formulas for the control limits, but process technicians are not expected to perform these calculations.

For a U chart, the *proportion* of counts is being charted. The average proportion is called $\overline{U}$. The standard deviation is similar to that of the C chart, but now we have to include a term (n) for the sample size. Figure 15.4 shows the top portion of a U chart.

ANY corp

Business unit: Isopropyl alcohol (worldwide ops)										
Month/Yr	01-20	02-20	03-20	04-20	05-20	06-20	07-20	08-20	09-20	10-20
No. defects	15	8	10	7	4	9	1	9	7	3
Sample size	600	500	550	725	350	450	550	475	625	325
No./sample	.025	.016	.0182	.0097	.0114	.02	.0018	.0189	.0112	.0092
UCL	.0232	.0244	.0237	.022	.027	.0251	.0237	.0247	.0229	.0277
LCL	0	0	0	0	0	0	0	0	0	0
Observation	1	2	3	4	5	6	7	8	9	10

Figure 15.4 Top portion of a U chart of ANY corp customer complaints from Figure 15.3.

Centerline (average) $= \overline{U}$

$$\text{UCL} = \overline{U} + 3\frac{\sqrt{\overline{U}}}{\sqrt{n}}$$

$$\text{LCL} = \overline{U} - 3\frac{\sqrt{\overline{U}}}{\sqrt{n}}$$

After plugging all the numbers into the equation, you come up with control limits that vary with each new month's data. Clearly, this is more complicated than any of us would care to manage on a routine basis, which is why this chart is not often used. It is technically correct, but it is complex to implement.

However, if this is the right chart to create, a statistical software package can be used to generate this chart. An example of a computer-generated U chart is shown in Figure 15.5.

Figure 15.5 Computer-generated U chart of customer complaints on a wide variety of samples sizes, based on data shown in Figure 15.4.

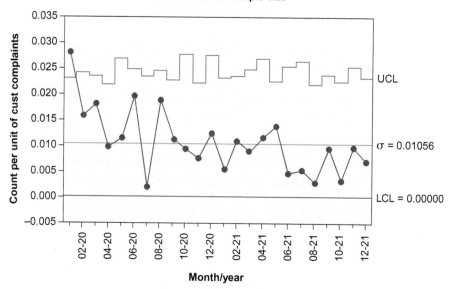

Figure 15.5 Computer-generated U chart of customer complaints on a wide variety of samples sizes, based on data shown in Figure 15.4.

15.3 Control Charts for Binomially Distributed Data

Let us say that you are working in the supply chain portion of a chemical plant. Your job is to fill drums with a certain quantity of chemical by weight, then seal and label the drum for shipment to the warehouse. Before you let the drums go out the door, you inspect them against a checklist such as this:

- Is the net weight of the drum within 1 kilogram (kg) of the target weight?
- Are the labels applied squarely and legibly on both the top and the side of the drum?
- Is the drum in good condition? (free of scratches, dents, rust, etc.)
- Are the drum bungs secured tightly to prevent leaks?

NP chart   control chart for counts of defectives (binomial distribution) with a constant sample size.

P chart   control chart for counts of defectives (binomial distribution) with a variable sample size.

If you note any deficiencies in the drums, then you must reject them and set them aside for rework. Shipping them to a customer would result in a customer complaint, which you are trying to avoid. In this case, the data are binomially distributed because there are only two options. The drums are either acceptable or unacceptable (rejected). You are not counting how many defects exist on each drum, but rather counting the number of defective drums. The appropriate control charts for these data would be the **NP chart** if the sample size is constant or the **P chart** if the sample size varies.

NP Charts

As mentioned earlier in this chapter, control charts for a constant sample size are easier to use than those for varying sample sizes, so let us start with the NP chart. The data for the NP chart are shown in Table 15.2.

The centerline, or average, $(n\bar{p})$ is calculated as follows:

$$\text{Centerline} = n\bar{p} = n\frac{\text{Total Defectives}}{\text{Total Inspected}}$$

The control limits are calculated using the following formulas:

$$\text{UCL} = n\bar{p} + 3\sqrt{n\bar{p}(1 - \bar{p})}$$

$$\text{LCL} = n\bar{p} - 3\sqrt{n\bar{p}(1 - \bar{p})}$$

So that you can see the results, the NP chart is shown in Figure 15.6.

Table 15.2 Data for NP Chart of Rejected Drums

Month/Year	Number of Defectives	Month/Year	Number of Defectives
01/20	7	01/21	5
02/20	1	02/21	2
03/20	4	03/21	3
04/20	1	04/21	5
05/20	3	05/21	1
06/20	1	06/21	0
07/20	3	07/21	1
08/20	4	08/21	1
09/20	2	09/21	1
10/20	1	10/21	1
11/20	0	11/21	3
12/20	1	12/21	0

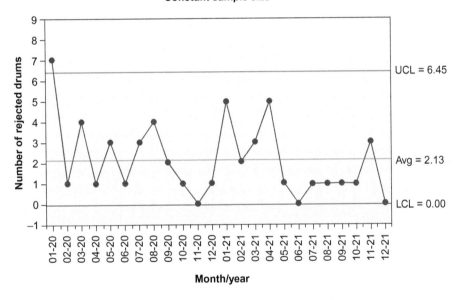

Figure 15.6 NP chart of the number of rejected drums, with data shown in Table 15.2.

P Charts

The NP chart of the binomial data shown in Figure 15.6 assumed that the sample size was constant. It did not even specify what the sample size was, but just said it was constant. If your sample size is not constant, then you must use the P chart instead. The data for the P chart are shown in Figure 15.7.

The number of drums that were filled each day varied greatly from day to day. The calculations for the P chart would be as follows:

$$\boxed{\text{Centerline} = \bar{p} = \frac{\text{Total Defectives}}{\text{Total Inspected}}}$$

And the control limits would be as follows:

$$\text{UCL} = \bar{p} + 3\sqrt{\frac{\bar{p}(1 - \bar{p})}{\sqrt{n}}}$$

$$\text{LCL} = \bar{p} - 3\sqrt{\frac{\bar{p}(1 - \bar{p})}{\sqrt{n}}}$$

Figure 15.7 Data for P chart of rejected drums.

ANY corp

Business unit:										
Month/Yr	01-20	02-20	03-20	04-20	05-20	06-20	07-20	08-20	09-20	10-20
No. defects	7	1	4	1	3	1	3	4	2	1
Sample size	600	500	550	725	350	450	550	475	625	325
No./sample	0.012	0.002	0.007	0.001	0.009	0.002	0.005	0.008	0.003	0.003
UCL	0.012	0.013	0.012	0.011	0.014	0.013	0.012	0.013	0.012	0.015
LCL	0	0	0	0	0	0	0	0	0	0
Observation	1	2	3	4	5	6	7	8	9	10

As was the case with the U chart, the P chart yields control limits that vary with each data point. The resulting control chart is shown in Figure 15.8.

Figure 15.8 P Chart of data shown in Figure 15.7. The image shows the proportion of rejected drums (see vertical label on the left side of this image). The data were collected from a wide range of sample sizes.

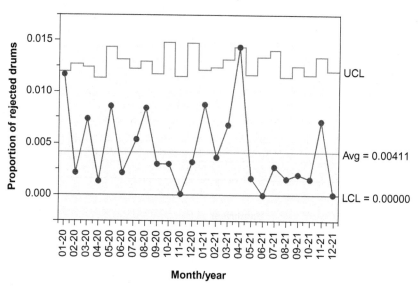

Once again, it is clear that this tool is difficult to implement. At this point, it is good to remember that how you define the problem determines which quality tools you will use. It is up to you to define the problem in such a way that you can study the situation simply and effectively.

Attributes Chart Selector

The next few pages examine the subject of sample size and tools. They compare several tools with one another to see what the impact on quality might be if you used the "wrong" tool or classified the data incorrectly. Figure 15.9 summarizes which attributes chart to use in which situation.

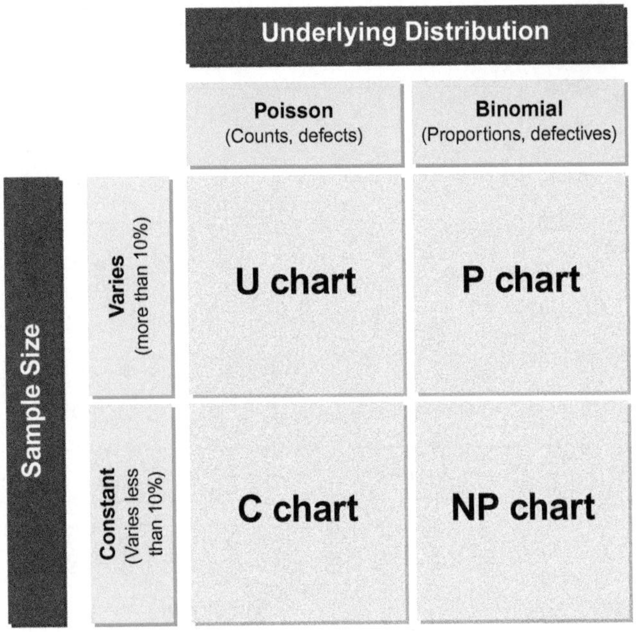

Figure 15.9 Diagram for selecting the appropriate attributes chart. Notice that the underlying distribution can either be defects or defectives (rejects), and the sample size can either vary or be constant.

Is the Sample Size Constant—or Close Enough?

The preceding sections have shown how to use different types of control charts for situations in which your sample size is constant (C and NP charts) versus when your sample size varies (U and P charts). Now let us see what happens when the sample size is "almost" constant.

The data from Figure 15.4 show the sample size varying from 325 to 725. This is a considerable range and resulted in control limits that varied noticeably in the U chart shown in Figure 15.5.

But what if the sample size data were much less variable? Would that make it easier? Consider the same counts of customer complaints but with a narrower distribution of shipments per month, which is what was used for the sample size (Figure 15.10, No./sample row). Now the control limits barely vary and yield a control chart that looks a little more like what you had been using (Figure 15.11).

You can see the same phenomenon when you look at control charts from the binomial distribution. The first data set had a wide variation in the sample size (see Figure 15.7). This resulted in control limits that varied noticeably in the P chart shown in Figure 15.8.

IN A NUTSHELL

If the sample size varies by 10 percent or less, you can consider the sample size to be "constant," allowing you to implement the simpler C or NP charts instead of the more complicated U or P charts.

Figure 15.10 U chart data with small variation in sample size.

ANY corp

Business unit: Isopropyl alcohol (world wide ops)									
Month/Yr	01-20	02-20	03-20	04-20	05-20	06-20	07-20	08-20	09-20	10-20
No. defects	15	8	10	7	4	9	1	9	7	3
Sample size	510	500	507	520	450	456	448	460	440	447
No./sample	0.0333	0.016	0.0197	0.0135	0.0089	0.0197	0.0022	0.0196	0.0159	0.0067
UCL	0.0242	0.0244	0.0243	0.0241	0.0251	0.025	0.0251	0.0249	0.0253	0.0251
LCL	0	0	0	0	0	0	0	0	0	0
Observation	1	2	3	4	5	6	7	8	9	10

Figure 15.11 U chart of data shown in Figure 15.10.

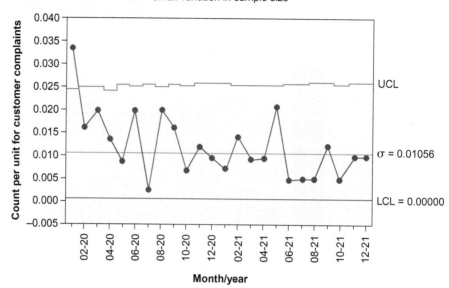

Consider the same number of defects but with a less variable sample size (Figure 15.12). The resulting P chart is shown in Figure 15.13.

Figure 15.12 P chart data with small variation in sample size.

ANY corp

Business unit: Isopropyl alcohol (world wide ops)										
Month/Yr	01-20	02-20	03-20	04-20	05-20	06-20	07-20	08-20	09-20	10-20
No. defects	7	1	4	1	3	1	3	4	2	1
Sample size	510	500	507	520	450	456	448	460	440	447
No./sample	0.012	0.002	0.007	0.001	0.009	0.002	0.005	0.008	0.003	0.003
UCL	0.012	0.013	0.012	0.011	0.014	0.013	0.012	0.013	0.012	0.015
LCL	0	0	0	0	0	0	0	0	0	0
Observation	1	2	3	4	5	6	7	8	9	10

Figure 15.13 P chart of data shown in Figure 15.12.

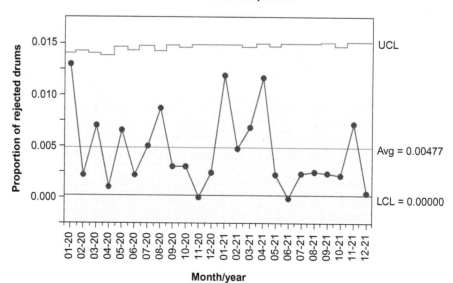

In both Figures 15.11 and 15.13, the control limits do not vary widely when the sample size is close to constant. For the sake of practicality, you can consider the sample size to be "constant" any time it varies by 10 percent or less. This allows you to use the simpler charts even when the application is not quite perfect.

It cannot be stated too often that your goal should be to understand the various tools in the quality toolkit so that the right tool can be applied to the right job at the right time to achieve the desired improvement. Sometimes more than one tool is available that will accomplish the job. The right tool may be a wrench—but you can choose between a crescent wrench, an open-ended wrench, a box wrench, or a socket wrench. If you choose a socket wrench, you may use a 6-point or a 12-point socket. Choosing the precise tool that is best for the job will make your work easier and more successful.

The quality field is filled with similar options. Process technicians need to choose wisely from the options provided. They also need to be practical about their choice and recognize that there are many tools that will accomplish the same end.

15.4 Interpreting Control Charts

Even though this section appears at the end of the chapter on attributes charts, it applies widely to variables and attributes charts, including x-bar, individuals, C, U, NP, and P charts.

Out of Control Data

Think back to the preceding discussions about variability. Recall that in a normal distribution, 99.7 percent of the data fall within 3 standard deviations of the average. By definition then, only 0.3 percent of the data fall outside these control limits. You have learned how to use control charts to plot data and detect signals of nonrandom behavior. So far, though, you have limited your signals to any points that fall outside the control limits.

Let us say you have a point on a graph. What are the odds that any given point will be below the average? There are two possibilities, and being below the average is one of the two. In statistical terms you would document the odds that any single point will be below the average as 50 percent.

$$\frac{1}{2} = 0.5 \quad (50\%)$$

What are the odds that a second point will be below the average? The second point is either above or below the average. Taking the second point all by itself, the odds are still 50/50. Now, what are the odds that *both* the first and the second points will be below the average? In other words, what are the odds of having two points in a row below the average? To determine these odds, take the odds that the first point will be below the average and multiply it by the odds that the second point will be below the average. The answer indicates that the odds of getting two points in a row below the average are 25 percent.

$$\frac{1}{2} \times \frac{1}{2} = \frac{1}{4} = 0.25 \quad (25\%)$$

Following this same logic, the odds of getting three points in a row below the average are 12½ percent.

$$\frac{1}{2} \times \frac{1}{2} \times \frac{1}{2} = \frac{1}{8} = 0.125 \quad \left(12\frac{1}{2}\%\right)$$

Extending this logic further, the odds of getting eight points in a row below the average are approximately 0.4 percent.

$$\frac{1}{2} \times \frac{1}{2} \times \frac{1}{2} \times \frac{1}{2} \times \frac{1}{2} \times \frac{1}{2} \times \frac{1}{2} \times \frac{1}{2} = \frac{1}{256} = 0.004 \quad (0.4\%)$$

This means that 99.6 percent of the time you see eight points in a row below the average, it would indicate that something is wrong. Only 0.4 percent of the time will eight points in a

Out of control points outside the three sigma (σ) control limits.

row below the average be "normal." The reason for selecting eight points in a row as a trigger to identify nonrandom behavior is that this closely matches the probability that a point will be outside the control limits. *The reaction triggers for points **out of control** have the same probability as the triggers for runs below the average.*

If you were to replace the words *below the average* with *above the average* in the preceding paragraphs, you would find that the math is the same. When you see eight points in a row above the average, 99.6 percent of the time this is also an indication that something is wrong. Only 0.4 percent of the time will eight points in a row above the average be "normal."

Shifts in Data

Shifts runs of eight points above or below the average.

Runs of eight points above or below the mean are called **shifts** in the data. Shifts are the second type of signal that something is wrong. They indicate that the process has moved from the expected average to some other value that is not expected.

A shift can be up or down from the expected average. Anytime a control chart yields a run of eight points above or below the expected average, it needs to be investigated to determine what has caused the process to shift. In Figure 15.14, there is a point that almost crossed the centerline, but it is still a below-average value, so this indicates a shift has occurred.

Figure 15.14 Example of a shift in data.

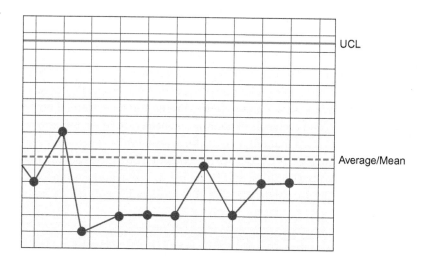

UCL

Average/Mean

Trends in Data

Finally, let us take a look at a similar situation. This time, instead of getting points in a row above or below the average, the process technician starts getting points that are trending up or down. Let us say you have a point on a graph. What are the odds that the next point will be above the first one? There are two possibilities; being above the previous point is one of the two (1/2). In statistical terms you would document the odds that any single point will be above the previous point as 50 percent.

What are the odds that a third point will be above the second point? The third point will be either above or below the second point. Taking the third point all by itself, the odds are still 50/50. Now, what are the odds that the second point will be above the first point and the third point will be above the second point? In other words, what are the odds that two points in a row will be trending up? Does this logic sound familiar? We can apply our earlier logic to extrapolate that the odds of getting eight points in a row trending up are approximately 0.4 percent. The odds of getting eight points in a row trending down are approximately 0.4 percent.

When you see eight points in a row constantly increasing or constantly decreasing, 99.6 percent of the time this is an indication that something is wrong. Only 0.4 percent of the time will eight points in a row moving up or moving down be "normal."

Trends in the data (shown in Figure 15.15) are the third type of signal that something is wrong. Any trend that has eight points in a row, whether increasing or decreasing, is not normal and should be investigated.

Trends   direction in a set of statistical data at a particular time; runs of eight points going up or going down.

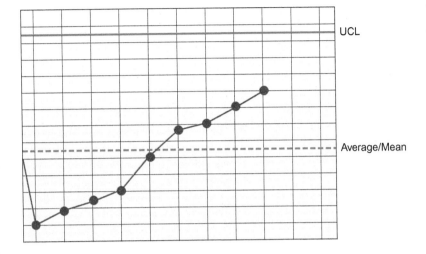

Figure 15.15 Example of a trend in data.

Three Main Interpretation Indicators

Three main rules for interpreting control charts follow:

1. Out of control—any points that exceed the three sigma control limits
2. Shifts—any run of eight points in a row below the average or above the average
3. Trends—any run of eight points in a row that constantly increase or decrease

After a control chart is set up and operational, process technicians should be looking for indications of special cause variation with each new point that is added to the chart. Abnormal values signal a need to investigate nonrandom behavior in the process.

This chapter has provided one simplified set of rules for general use. However, there are literally dozens of control chart interpretation rules available. If a company you work for has published its desired set of interpretation rules, that is what you will follow.

All of the various rules and variations of the rules exist for the same purpose—to determine when to react to the data and investigate potential problems in the process. If the cost of failures is extremely high in a certain business, then it should implement rules that have everyone react more quickly in order to avoid them.

Summary

Attributes control charts enable you to apply the most effective tools to your quality improvement efforts when your data are not normally distributed.

If the data are Poisson distributed (counts of defects, safety stats, or customer complaints), then use C and U charts depending on the sample size question—C charts for a constant sample size and U charts for a variable sample size.

If the data are binomially distributed (defectives), then use NP and P charts, again depending on whether the sample size is constant (NP) or variable (P).

If the sample size in the data varies by less than 10%, the sample size is considered to be constant.

As with all of the control charts examined in Chapter 14, *Variables Control Charts and Interpretation*, the purpose of attributes control charts is to distinguish common cause from special cause variation. Control charts are powerful tools for accomplishing this purpose. It is often more important to implement a control chart than it is to worry about exactly which chart is selected for use. Oftentimes different charts would indicate much the same thing.

When you detect special cause variation, you can investigate your process to determine the cause and remedy it. The three main indicators for interpreting control charts are:

1. Out of control data
2. Shifts in data
3. Trends in data

Computer software is available to generate the initial statistics and set up charts. The decision about use of computer software for ongoing plotting of control charts depends on the application of the control chart. Plotting charts by hand also has value and should be considered when making decisions.

Checking Your Knowledge

1. Define the following key terms:
   a. Attributes data
   b. C chart
   c. NP chart
   d. Out of control
   e. P chart
   f. Shifts
   g. Trends
   h. U chart
   i. Variables data

2. C charts can be used to plot what kind of data?
   a. Measures of product quality such as purity and pH
   b. Counts of customer returns
   c. Proportions of rejected drums
   d. Quality audit compliance statistics

3. NP charts can be used to plot what kind of data?
   a. Measures of product quality such as purity and pH
   b. Counts of customer complaints
   c. Proportions of defective drums
   d. Quality audit compliance statistics

4. For a Poisson distribution and in a system with a constant sample size, the most correct chart to employ would be:
   a. C
   b. U
   c. NP
   d. P

5. For a Poisson distribution and in a system with a variable sample size, the most correct chart to employ would be:
   a. C
   b. U
   c. NP
   d. P

6. For a binomial distribution and in a system with a constant sample size, the most correct chart to employ would be:
   a. C
   b. U
   c. NP
   d. P

7. For a binomial distribution and in a system with a variable sample size, the most correct chart to employ would be:
   a. C
   b. U
   c. NP
   d. P

8. It is acceptable to assume sample size is constant enough, and so to employ the simpler control charts, if the sample size varies by no more than:
   a. 5 percent
   b. 10 percent
   c. 15 percent
   d. 20 percent

9. There are three main rules for interpreting control charts. (Select all that apply.)
   a. Out of control
   b. Runs
   c. Shifts
   d. Trends

10. The use of a constant sample size simplifies the control charts because:
    a. variation is bad, so you like to use only numbers that are constant.
    b. constant sample size allows you to have only one set of control limits.
    c. constant sample size does not simplify control charts.
    d. variation in the sample size causes you to react too quickly to out of control data.

11. Match the following rules for interpreting control charts with their meanings:

Rules	Definition
I. Out of control	a. Runs of eight points going up or going down
II. Shifts	b. Points outside the three sigma control limits
III. Trends	c. Runs of eight points above or below the average

12. (*True or False*) Poisson distributions are an example of attributes data.

13. (*True or False*) Normal distributions are an example of attributes data.

14. (*True or False*) Binomial distributions are an example of attributes data.

15. (*True or False*) Control charting should always be done by computer software, never by hand.

NOTE: Answers to Checking Your Knowledge questions are in the Appendix

Student Activities

1. Examine the chart in Figure 15.3. Apply the three main indicators for interpreting control charts. What point or points would you investigate? What do the last seven points on this graph tell you?

2. Given the following set of customer complaint data, plot a C chart. Calculate the average ($\overline{C}$) and the control limits using the control limit formulas. Are all of these points "in control"? Do you see any evidence of special causes?

Month/Year	Number of Defects	Month/Year	Number of Defects
01/20	3	01/21	2
02/20	0	02/21	0
03/20	3	03/21	3
04/20	3	04/21	3
05/20	2	05/21	0

Month/Year	Number of Defects	Month/Year	Number of Defects
06/20	1	06/21	0
07/20	2	07/21	0
08/20	3	08/21	0
09/20	1	09/21	5
10/20	0	10/21	0
11/20	2	11/21	1
12/06	2	12/21	2

Chapter 16
Process Capability

"Data will talk to you if you're willing to listen."

~ JIM BERGESON

Objectives

Upon completion of this chapter you will be able to:

16.1 Interpret process data and explain how they relate to the needs of the customer. (NAPTA Quality, Capability 1, 8*) p. 225

16.2 Calculate key process capability indices (Cp, Cpk, and Ppk, and so on) and explain what each index is telling you. (NAPTA Quality, Capability 4, 5, 6, 7) p. 225

16.3 List and explain potential sources of process variability. (NAPTA Quality, Capability 8) p. 232

16.4 Explain the concept of measurement system variation and the importance of using the right tool for the job. (NAPTA Quality, Data Collection 8) p. 233

16.5 Explain the concepts of overcontrol and their impact on process capability. (NAPTA Quality, Capability 3) p. 234

Key Terms

Lower specification limit (LSL)—the lowest level allowed by the customer for a specific quality measure, **p. 226**

Measurement capability index (Cm)—the ratio of your specification band to your measurement process variability, **p. 233**

Process capability index (Cpk)—the ratio of the customer's specification range to your process variability taking the average (process center) into account; this index assumes the data have been "cleansed" of special causes, **p. 228**

---

*North American Process Technology Alliance (NAPTA) developed curriculum to ensure that Process Technology courses will produce knowledgeable graduates to become entry-level employees in process technology. Objectives from that curriculum are named here in abbreviated form. For example, "(NAPTA Quality, Capability 1, 8)" means that this chapter's objective 1 relates to objectives 1 and 8 of the NAPTA curriculum on capability.

Process performance index (Ppk)—the same calculation as the Cpk but without the assumption that the data have been cleansed of special causes, **p. 231**

Process potential index (Cp)—the simple ratio of the customer's specification range to normal process variability without taking the average into account, **p. 226**

Process potential C_{pl}—process potential for processes that have only a lower specification limit, **p. 231**

Process potential C_{pu}—process potential for processes that have only an upper specification limit, **p. 231**

Upper specification limit (USL)—the highest level allowed by the customer for a specific quality measure, **p. 226**

16.1 Introduction

The clear fact is that processes vary. We have shown how to display the variability of a process graphically with a histogram. We have shown how to measure that variation mathematically and how to describe the distribution of data by using the mean to define the center of the distribution and the standard deviation, or sigma, to define the spread. Customers desire consistency in the products they buy, and they object to variation.

This chapter will show how to relate the distribution of data to the needs of the customer. Completely eliminating all variation from processes and product is not possible. The question is not, "How do I get rid of all variation?" Instead, the question is, "How much variation can the customer accept?"

There is a practical limit to the ability to measure variation. There is also a certain amount of variation that customers can tolerate without a negative impact on their process. This chapter shows how to calculate several indices to help judge the amount of process variability compared to what the customer needs. These include process potential index (Cp), process capability index (Cpk), and process performance index (Ppk). Measurement variation is such an important topic that it has its own index, called the measurement capability index (Cm).

Once the acceptability of a process's variation is determined, the next step is to examine potential sources of variation. For example, people on different shifts may have different ways of accomplishing the same task. This can lead to an increase in variation. Since all processes vary, it is logical that the process of taking a sample, the process of analyzing a product, and even the process of storing the product may vary.

This chapter looks at potential sources of variability in more detail and explains the importance of understanding the contribution each one makes to the total variability that the customer sees. One additional aspect of variability that will be explored is the concept of overcontrol.

Learning how to design these experiments and analyze the data is part of a more advanced course. However, it is important to know that the tool exists, that it is used, and that process technicians gather the data. Process technicians must understand what the tool does and how it helps improve processes. See Chapter 12, *Data Collection and Representative Sampling*.

16.2 Key Process Capability Indices

As discussed in Chapter 13, *Variance and Operating Consistency*, in a normal distribution 99.7 percent of all the data are expected to be within 3 standard deviations (sigma, or σ) of the mean. With a 3 sigma range above the mean and a 3 sigma range below the mean, the total spread of the distribution is 6 standard deviations wide. Six times sigma is how much variability there is in a process with normal distribution.

Upper specification limit (USL)
the highest level allowed by the customer for a specific quality measure.

Lower specification limit (LSL)
the lowest level allowed by the customer for a specific quality measure.

Now we want to relate that variability to what is important to the customer. Most variables that are important to the customer will have a range of acceptability defined, typically called the *tolerance, specification* (spec), or *specification range*. The upper end of the specification range is referred to as the **upper specification limit (USL)**. The lower end of the specification range is referred to as the **lower specification limit (LSL)**. These numbers define how much variability the customer is willing to accept.

Keep in mind that just because a customer *accepts* this much variability, the customer may not *want* this much variability. One way of comparing a company's process variability to the customer's needs is to add the customer's specification limits to a histogram (Figure 16.1). This quick visual will tell you and the customer whether the company can expect its process to meet the customer's needs. In this figure, the +3 sigma and −3 sigma limits represent what is normal for the process. The USL and LSL limits indicate what the customer is willing to tolerate. Because the process falls well within the customer's tolerance limits, it seems to be acceptable. However, there is a simpler way of determining whether a process is acceptable.

Figure 16.1 Example of the relationship of process spread to specification limits.

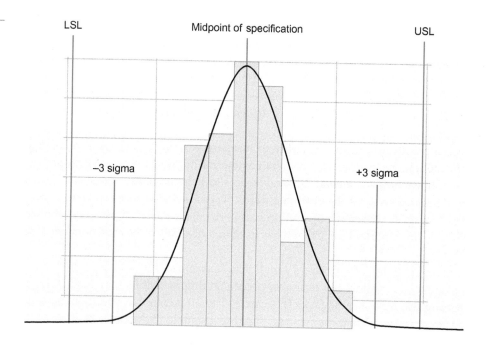

Process Potential Index—Cp

The process potential index (Cp) was developed as a mechanism to provide a single number that would indicate whether a process is capable of meeting the customer's needs. The **process potential index (Cp)** is a simple ratio of the customer's specification range to the company's normal process variability range. With the normal process variability being defined as ±3 sigma, or 6 times sigma, the formula is as follows:

Process potential index (Cp)
the simple ratio of the customer's specification range to normal process variability without taking the average into account.

$$Cp = \frac{USL - LSL}{6 \times \sigma}$$

Logically, we want the process variability (the denominator) to be small compared to the tolerances (the numerator). If process variability is greater than the customer's needs, it will result in a Cp index of less than 1.0. This result indicates that some percentage of the product will not meet the specifications.

Figure 16.2 shows an example of a process. The specifications are set at 42 to 58 with a target, or nominal, value of 50. This process is centered around the nominal value but has a calculated standard deviation (σ) of 3.33, which means the process varies ±10 (3 × σ), but the customer desires a range of ±8.

Figure 16.2 Example of a centered process that is not capable of meeting customer specifications.

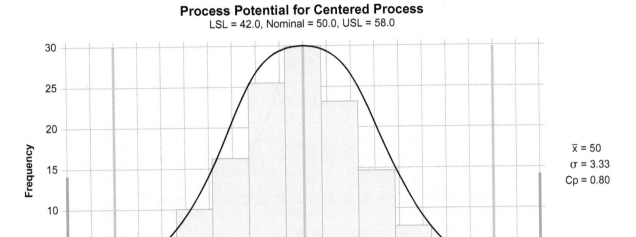

By replacing the terms in the formula (below) with the numbers provided, a Cp index of 0.8 is obtained. Since it is less than 1.0, this index indicates that, a percentage of the time, the process will produce product that does not meet the needs of the customer.

$$Cp = \frac{USL - LSL}{6 \times \sigma} = \frac{58 - 42}{6 \times 3.33} = \frac{16}{19.98} = 0.8$$

If the top and bottom of the equation were the same, then it would give a Cp index of 1.0. This would mean that the product would normally vary within the boundaries of the customers' needs 99.7 percent of the time—which is acceptable for a lot of processes.

Although a Cp index of 1.0 is good (Figure 16.3), it does not provide any safety net. It is preferable to have a process that fits well inside the customer requirements.

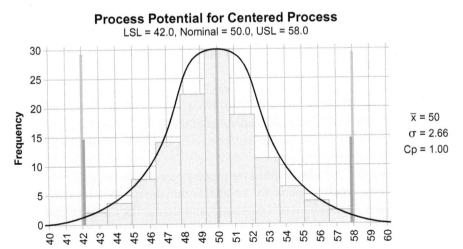

Figure 16.3 Example of a centered process that is marginally capable of meeting customer specifications but has no safety net.

If process variability is less than the customer's needs, it will result in a Cp index greater than 1.0 (Figure 16.4). This indicates that the process will virtually always make product that passes specifications, which is what you want.

Figure 16.4 Example of a centered process that is capable of meeting customer specifications with a safety distance on both sides.

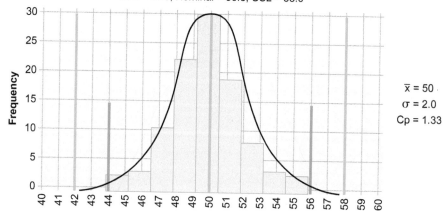

Process Potential for Centered Process
LSL = 42.0, Nominal = 50.0, USL = 58.0

$\bar{x} = 50$
$\sigma = 2.0$
Cp = 1.33

From the customers' perspective, the higher the Cp, the better they like it. The lower the Cp, the more variability the process exhibits in relation to their tolerances, and the less they will like it. Customers want a product to use the minimum amount of their available specification range. By inverting the Cp and multiplying by 100, it is even possible to calculate what percentage of the specification range the process takes up at different Cp indices (shown below).

- A Cp of 2.00 (6 sigma) means the six sigma is 50 percent of the specification range.
- A Cp of 1.66 (5 sigma) means the six sigma is 60 percent of the specification range.
- A Cp of 1.33 (4 sigma) means the six sigma is 75 percent of the specification range.
- A Cp of 1.00 (3 sigma) means the six sigma is 100 percent of the specification range.
- A Cp of 0.66 (2 sigma) means the six sigma is 150 percent of the specification range.
- A Cp of 0.33 (1 sigma) means the six sigma is 300 percent of the specification range.

The problem with using this index is that it tells only what a process has the potential of doing if it is perfectly centered within the specifications. Looking back to Figure 16.4 we see a process with a total specification range of 16 (USL − LSL = 58 − 42 = 16). The process variability is 6 times the sigma, which gives you a process spread of 12 (6 × σ = 6 × 2 = 12).

Using the formula for the Cp, the Cp index is calculated as 16 ÷ 12 = 1.33. Normal process variability is much less than the specifications. In this particular case, there is a safety net of 1 standard deviation on each side of the process between the process and the specifications. A good Cp index indicates that the process is capable of meeting specifications as long it is running in the center of the range—in other words, process variability is within acceptable limits.

In Figure 16.5, there are the same numbers for specification limits and the same process sigma, so the result is the same Cp of 1.33. What is the problem? In this case, the process is not centered at the middle or nominal value of the specification, which is 50. Instead, the process is running at an average of 54. Because of this, some product will fail to meet the customer requirements.

The reason the Cp index is called the process potential index is that it tells how capable a process is if the process is perfectly centered within the specification range. Recall the two key measures of distributions of data—mean and sigma. The Cp index uses only the sigma in its judgment of acceptability, so it can only tell whether the spread of the data is good enough. It cannot say whether the process is actually on target.

Process Capability Index—Cpk

Process capability index (Cpk)
the ratio of the customer's specification range to your process variability taking the average (process center) into account; this index assumes the data have been "cleansed" of special causes.

The **process capability index (Cpk)** is the ratio of the customer's specification range to process variability, taking into account where the process is running in relation to the

Figure 16.5 Example of an uncentered process that is not capable of meeting customer specifications.

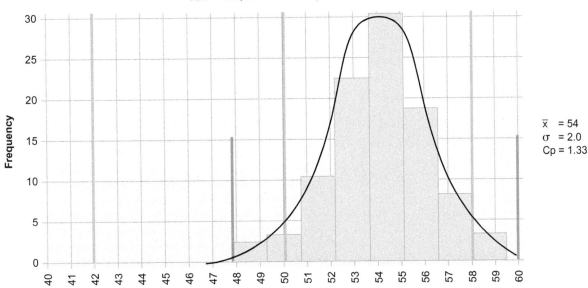

These values indicate that there is a process capability of 0.67 against the upper specification limit and a process capability of 2.0 against the lower specification limit. The interpretation of the Cpk index is the same as for the Cp index. A value of 1.0 is good. Values above 1.0 are better. Values below 1.0 are not good. We can conclude that this process is capable of meeting the lower specification limit but not capable of meeting the upper specification limit. The value that is reported is the worse of the two values, so this Cpk would be reported as 0.67 (Figure 16.6).

specification range. If the process is not centered, then it stands to reason that the process is running either to the left or to the right of the middle. The Cpk index is calculated by comparing the process to both the upper and the lower sides of the specification, then reporting the worst-case scenario. Here are the formulas:

$$\text{Cpk(upper)} = \frac{\text{USL} - \bar{x}}{3 \times \sigma} \quad \text{or} \quad \text{Cpk(lower)} = \frac{\bar{x} - \text{LSL}}{3 \times \sigma}$$

The Cpk that is reported is the minimum value of these two calculations. Let us try it out for Figure 16.5:

$$\text{Cpk(upper)} = \frac{58 - 54}{3 \times 2} = \frac{4}{6} = 0.67 \quad \text{Cpk(lower)} = \frac{54 - 42}{3 \times 2} = \frac{12}{6} = 2.0$$

These values indicate that there is a process capability of 0.67 against the upper specification limit and a process capability of 2.0 against the lower specification limit. The interpretation of the Cpk index is the same as for the Cp index. A value of 1.0 is good. Values above 1.0 are better. Values below 1.0 are not good. We can conclude that this process is capable of meeting the lower specification limit but not capable of meeting the upper specification limit. The value that is reported is the worse of the two values, so this Cpk would be reported as 0.67 (Figure 16.6).

If the Cpk and the Cp calculations give exactly the same index, this indicates that a process is centered within the specifications. The Cp will always be equal to or greater than the Cpk because the Cp represents the best the Cpk can be if the process is perfectly centered. In this case, the Cpk is considerably lower than the Cp. It is clear that the process is not centered. By looking at the Cpk calculations we know that the process is off center to the high side ($\bar{x} = 54$, while nominal or desired $= 50$). We also know that the total variation of the process is acceptable (because the Cp is greater than 1.0). If the process can be shifted down from 54 to 50, then it will be capable of meeting all the customer's specifications all the time.

A Cpk of 1.0 (Figure 16.7) would indicate that your normal process variability, defined as $\pm 3\ \sigma$ from the mean, was colliding with one of the specification limits. If the Cp is greater than 1.0 for this process, then you know you have to shift the process to the middle of the target range to achieve a capable process.

Figure 16.6 Example of an uncentered process that is not capable of meeting customer specifications.

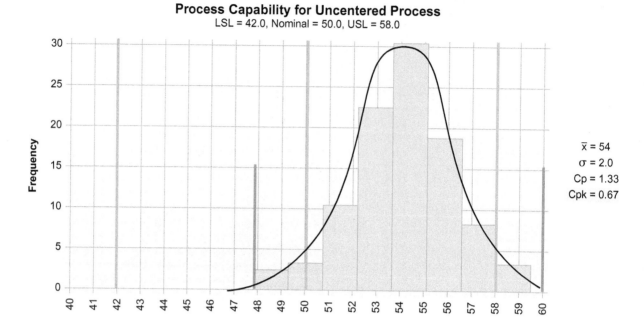

Figure 16.7 Example of an uncentered process marginally capable of meeting customer specifications.

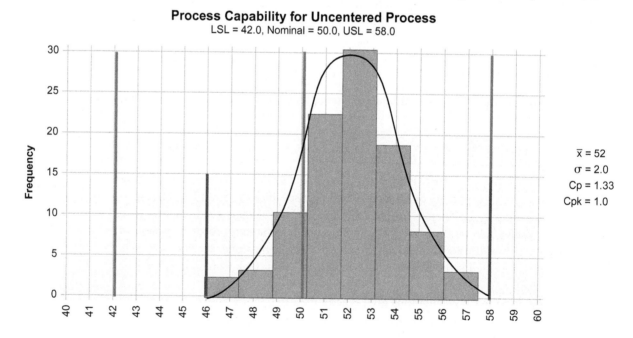

What customers want to see is a process in which the Cpk is well above 1.0, as shown in Figure 16.8. A Cpk index of 1.33 or greater is a common target in the process industries. Mathematically, this means that the specification limits are equal to 4 standard deviations of the process. As noted, though, the math is less important than just knowing that a result above 1.0 is required and a result above 1.33 is desired.

The Cpk index assumes that a process is stable and that there were no assignable causes in the data.(See Chapter 13, *Variance and Operating Consistency*, for review of assignable causes.)

Figure 16.8 Example of an uncentered process that is capable of meeting customer specifications.

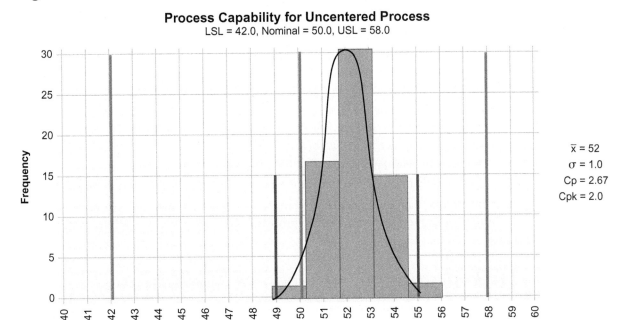

Process potential C_{pu} is a measure of the potential capability of the process based on its upper specification limit. C_{pu} is a ratio that compares two values:

- The distance from the process mean to the upper specification limit (USL)
- The one-sided spread of the process (the 3-σ variation) based on the variation within the subgroups.

In contrast, **process potential C_{pl}** is the process capability based on the lower specification limit.

Process Performance Index—Ppk

Ideally, a process is stable and there are no assignable causes in the data. Many times, however, process technicians are compelled to use the data they have, even though they know that there were "issues" with the process during the collection of the data. For this purpose, there is the **process performance index (Ppk)**.

Many software packages provide an option to generate a Ppk index instead of a Cpk index. The process performance index (Ppk) calculations do not assume the data have been cleansed of special causes. When the Ppk index is reported, it indicates that whatever data were available were used, and those data may not be completely trustworthy. The data do not meet the rules of a process that is in control. For example, the data included are above or below the control limits.

When process technicians report a Cpk index, they are saying that the data have been cleansed (the process is in a statistical state of control) of all assignable (special) causes and that the index truly reflects what the process is capable of under the best of circumstances. Again, there is no difference in calculations and no difference in interpretation of the results. There is only a difference in the assumptions about the integrity of the data.

Capability Index Summary

1. The Cp index is the process potential index. It is the maximum a Cpk index can achieve. It does not take into account where a process is running (centered versus off center). Consequently, it can be used only to judge how good the spread of a process is relative to the spread of the customer's specifications.

Process potential C_{pu} process potential for processes that have only an upper specification limit.

Process potential C_{pl} process potential for processes that have only a lower specification limit.

Process performance index (Ppk) the same calculation as the Cpk but without the assumption that the data have been cleansed of special causes.

2. The Cpk index is the process capability index. It takes into account where a process is running in addition to how much spread there is in a process compared to the customer's specifications. The Cpk index assumes that data have been cleansed of assignable causes and represent the capability of the process.

3. The Ppk index is the process performance index. It is calculated the same as the Cpk but assumes that the data have *not* been cleansed of assignable causes and that the index is a reflection of how a process is currently performing, not what it is capable of performing.

4. For all three of these indices, values of 1.0 or greater are desired, with 1.33 being the industry standard target. Values below 1.0 are not desired because they indicate that the processes are not capable of meeting the customer's needs.

5. For a perfectly centered process, the Cp and the Cpk will be the same number.

6. For a process that is not centered within the specification range, the Cp is the maximum number that the Cpk could be if the process were to be shifted toward the center.

16.3 Potential Sources of Variation

Once we know how to tell whether a process is capable of meeting the needs of the customer, we need to put controls in place to monitor the process and ensure it stays capable and in control. For this we use control charts, tools employed to monitor processes for changes. The preceding chapters talked about the various types of control charts that are available.

But what do we do if a process is not capable of meeting the needs of the customer? Let us assume that the process is stable, which means there are no special or assignable causes (just random variation), but the process variability either is not centered within the specification range or is too wide compared to the specification range. A process that is not centered within the range leads to an easy conclusion: just make an adjustment and shift the process to the center of the range. But what if the variation is too wide? What can be done about a process that varies more than the customer can tolerate? The obvious answer is to attempt to reduce the variability, but how? Where should efforts to reduce variability be applied? Some potential sources of variability may include the following:

- Shift-to-shift differences in operation
- Changes in raw material lots
- Seasonal variation (cooling water temperature can vary greatly from summer to winter, yielding different cooling capacities in a unit)
- Top of the storage tank to bottom of the storage tank differences
- Sampling variability
- Measurement system variability.

Think of the total process variation that the customer sees as a block of variation (Figure 16.9). There are many components to this block of variation, and they all add up to a number that the customer cannot tolerate. The technician's job is to dissect this block into its individual components. Then it will be possible to see where to spend time making improvements. Process technicians can choose to work on the easiest things to fix or on the things that add the most variation to the big block. Decisions about what to do depend on how much the improvement will cost and how much of an improvement it can produce.

Another factor to consider is that sometimes the variation overlaps. In other words, you cannot just add up the standard deviations of the components and get the standard deviation for the total. Conceptually, this is exactly what we do, but mathematically it takes a little more work. The tools for segregating total variation are beyond the scope of this textbook.

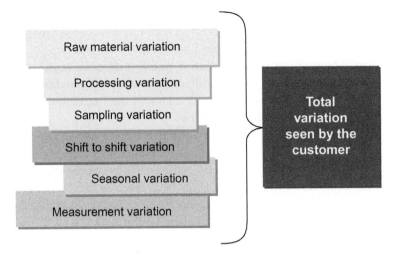

Figure 16.9 Partial sources of variation and total variation seen by customers.

Raw material variation

Processing variation

Sampling variation

Shift to shift variation

Seasonal variation

Measurement variation

Total variation seen by the customer

16.4 Measurement Capability Index—Cm

The process of analyzing a sample has variation just like the process for making the chemical in the first place. Many of the same components of variation can exist within the measurement process, so we will briefly discuss measurement system variation.

A measurement process may result in variation in many ways. A common example of this variation is that different lab analysts running the samples on different days can get different results (Figure 16.10). Many chemical analyses use chemical reagents. A change from one bottle of reagent to another can alter results. Laboratory instrumentation can "drift" off calibration just like process equipment instrumentation can.

Figure 16.10 Lab analysis results can show variation related to small changes in the laboratory.

A.

B.

For these reasons, the analytical process must undergo the same kind of scrutiny that the main process undergoes. The analytical process might be capable of meeting the needs of the process. A thorough understanding of analytical variability is a crucial part of understanding the process variability. The capability of an analytical process has its own capability index, the **measurement capability index (Cm)**.

The calculations for the Cm will look familiar. Here is the equation:

$$Cm = \frac{USL - LSL}{6 \times \sigma}$$

Measurement capability index (Cm)   the ratio of your specification band to your measurement process variability.

It is the same formula as for the process potential index or Cp. The only difference is that instead of using the sigma (σ) for the entire process, we use the sigma for the measurement process only. This is accomplished by taking the same sample of material and analyzing it 15 to 20 times. When the same analyst measures the same sample using the same equipment and the same reagents and gets different numbers, the standard deviation for the analytical method can be calculated. This eliminates all the variation due to anything related to the process of manufacturing the product.

The interpretation of the Cm differs from that of the Cp. Recall from the earlier discussion that a Cp of 1.0 means that the variation of the process is equal to what the customer needs, or the specifications. If you applied that same thinking here, you might conclude that a Cm of 1.0 means that the variation of the measurement process is equal to the specifications. However, that conclusion would be incorrect because the measurement process is just one of many components of the total variability.

From a practical standpoint, if a measurement system varies within the entire range of the specifications, no variation would be allowed in the rest of the process. A Cm of 1.0 or less means that when you get a result back from the lab, it could be on the high side of the specification, the low side of the specification, or anywhere in between, and yet nothing has changed in the process.

Many companies in the processing industries have accepted a minimum Cm target of 5.0. This is equivalent to the measurement process's consuming 20 percent of the total specification range. Some Six Sigma work targets set the goal as high as 10.0.

The higher the Cm, the less variability the measurement process has compared to the specification range, and the greater the chances that total variability will be acceptable. The lower the Cm, the more variability the measurement process has compared to the specification range.

It may not be the job of process technicians to analyze samples or determine the method capability. However, if they know how capable analytical methods are, they can know when to react to lab results. If the analytical process is capable (meaning it has a Cm of at least 5.0), then the process technician moves on to other sources of variability. But what if it is not?

Sometimes the issue is that the measurement tool being used may not be right for the task. For example, using a standard ruler to measure the length of a stick to the nearest thousandth of an inch simply will not work. In this case, what is really needed is a micrometer. Similarly, to know the pH of a chemical to the closest whole number, litmus paper may be a good enough tool, but to measure it to the second decimal place probably requires a sophisticated analytical technique.

As is the case with quality in general, success comes with knowing what is needed and in using the right tool for the right job at the right time. It is important to know how variable the measurement process is and how much variability can be tolerated, in order to select the right tools.

16.5 Overcontrol

This text has covered variation and has established that the total variation can come from a wide variety of sources. Sometimes the sources of the variation are part of the process (normal variation), and sometimes they are episodic (abnormal variation).

In the previous chapters, the topic of control charts was covered. The purpose of control charts is to distinguish between common cause and special cause variation, in order to know when to react and when not to react. Without the appropriate knowledge of what is happening in a process, people might react to numbers when they should not. Reacting to common cause variation actually adds more variation into the process.

When the process capability of a process is being overcontrolled, the process will naturally be less capable. Dr. Deming would walk into a manufacturing plant and tell all the workers to take their hands off the equipment so he could see what the equipment could do on its own. He documented an average 40 percent reduction in variation just by eliminating the overcontrol that is so common in industry.

The concept of overcontrol can be demonstrated in the classroom using a *quincunx* (Figure 16.11). A quincunx or "Galton board" is a triangular array of pegs. Balls are dropped onto the top peg and bounce down to the bottom where they are collected in little bins. Each time a ball hits one of the pegs, it bounces either left or right. In the end, the balls form a normal distribution curve.

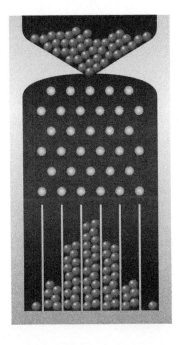

Figure 16.11 Quincunx.
CREDIT: Peter Hermes Furian/Shutterstock.

Deming also played an interesting game with managers, called the "red bead experiment." In this experiment, Deming used a bowl with red and white beads to demonstrate the futility of reacting to common cause variation (in this case, a high percent of red beads among the white beads that were supposed to be produced). When managers pulled scoops of beads out of the bead bowl, they could not consistently avoid getting red beads too. The red bead experiment shows managers that if a system has too much variation in it, no amount of effort by the staff will be able to eliminate the variation.

Let us say the process variability has been well documented, and you know that the process varies ±3 from the target value when all is running well. This ±3 variation includes process and analytical variation. Getting any value within this range means the equipment should be left alone because it is doing what it is capable of doing. If, on the other hand, you get a value that is 2 units lower than target, and you adjust your process up by 2 units to compensate, the entire process spread of 6 units (±3) will no longer be centered. The probability of getting a higher number increases.

When that higher number appears, let us say you adjust the process down to compensate again. In the end this creates a total process, made up of several smaller ones, that was generated by moving the process around manually. The total process capability is greatly reduced, and the variability of the product is greatly increased—much to the dissatisfaction of customers. Figure 16.12 shows how one big histogram may in fact be made up of several smaller histograms that process technicians create by not understanding the basic concepts of variation.

Figure 16.12 Multiple subprocesses make up the grand process.

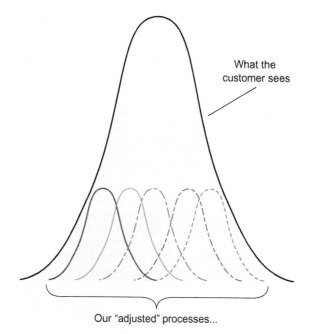

What the customer sees

Our "adjusted" processes...

Failure to understand the measurement system variation can lead to overcontrol of the process, which adds even more variability to what is inherent in the process. Consider a common setting. Did you ever see someone set the air conditioning thermostat to 65 degrees because the room was too hot? Does this make the air conditioner run any cooler? No. Either the AC blows cold air or it does not. Setting the thermostat down a degree or two will cause it to come on and cool off the room a degree or two. Setting the thermostat down lower does not make the room cool off any faster. It just eventually makes the room get too cold and overcontrols the temperature.

Summary

Knowing that processes vary is a good start. Relating that variation to the needs of the customer is an important next step. The process potential index (Cp), process capability index (Cpk), and process performance index (Ppk) provide simple numerical indices to help students understand the relationship between the variability of the processes and the needs of the customer.

Several concepts must be understood in order to make proper use of knowledge about process capability:

- Variation can come from many different sources. Some examples include seasonal variation, shift-to-shift variation, sampling variation, and measurement system variation.

- All capability indexes are calculated using measures from the process or products, so understanding the amount of variation that comes from the measurement process is critical to all improvement efforts. The measurement capability index (Cm) provides a simple numeric evaluation of the adequacy of the measurement system.

- Statistical tools are employed in process industries to help identify where to apply improvement efforts.

- The sensitivity of the measurement device must be adequate for the measurement needed. Sometimes a standard ruler will do; some situations require measuring down to a micrometer.

Failure to understand the measurement system variation can lead to overcontrol of the process. Overcontrol, which introduces small variations, ultimately adds more variability to the process.

Checking Your Knowledge

1. Define the following key terms:
   a. Lower specification limit (LSL)
   b. Measurement capability index (Cm)
   c. Process capability index (Cpk)
   d. Process performance index (Ppk)
   e. Process potential index (Cp)
   f. Process potential C_{pl}
   g. Process potential C_{pu}
   h. Upper specification limit (USL)

2. You have been working on understanding your process variability and how this variation might be perceived by your customers. Given the following information, what is your Cp?

$\bar{x} = 102$
$\sigma = 2$
USL $= 106$
LSL $= 94$

   a. 0.8
   b. 0.9
   c. 1.0
   d. 1.1

3. Using the same information from question 2, what is your Cpk?
   a. 0.67
   b. 0.76
   c. 1.0
   d. 0.5

4. Again using the same information from question 2, what is your Ppk?
   a. 0.67
   b. 0.76
   c. 1.0
   d. 0.5

5. Given a specification range of 95 to 110 and a measurement system standard deviation of 0.5, what is your Cm?
   a. 1.0
   b. 3.0
   c. 5.0
   d. 7.0

6. What is the minimum Cm to ensure that the measurement system is adequate?
   a. 1.0
   b. 3.0
   c. 5.0
   d. 7.0

7. (*True or False*) The process potential index and the measurement capability index use exactly the same calculations.

8. Which of the following would not be listed as a source of variability?
   a. Shift-to-shift differences in operation
   b. Changes in raw material lots
   c. Sampling variability
   d. Changing your PPE
   e. Measurement system variability

9. For a process to be capable of meeting the customers' needs, it would need a capability index of:
   a. Less than 1.0
   b. Exactly 1.0
   c. Greater than 1.0
   d. Zero

10. (*True or False*) Measurement system variation does not affect the total variation in the process.

11. What happens to process variability when the technician adjusts the process in response to common cause variation?
    a. The variability decreases.
    b. The variability increases.
    c. There is no change in the process variability.

12. (*True or False*) Any measurement device will do, as long as it gives a number.

13. A Cpk index of _____ or greater is a common target in the processing industries.
    a. 0.67
    b. 1.00
    c. 1.33
    d. 3.00

14. (*True or False*) After identifying sources of variability in your process, you would choose the most difficult problems that add the least variability to work on first.

15. Which of the following is not usually the role of the process technician?

    a. Take samples.

    b. Label samples.

    c. Analyze data.

    d. Collect data.

16. Deming found that by eliminating overcontrol there was an average reduction in variation of how much?

    a. 10 percent

    b. 20 percent

    c. 30 percent

    d. 40 percent

NOTE: Answers to Checking Your Knowledge questions are in the Appendix.

Student Activities

1. List the strengths and weaknesses of a Cpk index versus a Cp index.

2. Compare and contrast the differences in the calculation of a Cpk index, a Cp index, and a Ppk index.

3. Why would you use a Cpk index over a Cp index? Explain.

Chapter 17
Putting the Puzzle Together

"The intelligence of the team exceeds the intelligence of the individuals in the team."

~ SENGE

Putting the Puzzle Pieces Together

You have come a long way in your quality journey. When you started down this pathway, you did not know which way to go and were not sure how you would get there (Figure 17.1).

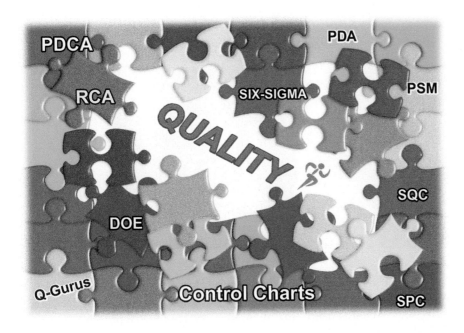

Figure 17.1 Some pieces of the Quality puzzle.

CREDIT: Courtesy of Willie L. Myles.

Now you have a firm foundation regarding the history and importance of the quality movement in the United States, as well as a working familiarity with the basic quality tools employed.

You now understand that the topics covered in the various chapters of this book all point toward a single goal—to make a consistent quality product that meets or exceeds the needs of customers so that they continue to purchase your products, keep your company in business, and keep you gainfully employed.

Introduction to Process Quality

In Chapter 1, you began your quality journey by learning a little of the history of the quality movement, starting out in Japan after World War II and gaining momentum in the United States in the late 20th and early 21st centuries. You gained an appreciation for how well established the quality concepts are today. You learned that few, if any, industries do not practice some form of quality improvement (Figure 17.2).

Figure 17.2 Control chart sample.

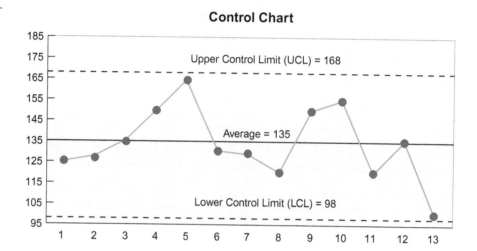

Quality is well entrenched in the daily activities of the process industries. Process technicians play a vital role in the success of the quality effort (Figure 17.3). Indeed, without the participation of the process technicians, no quality effort can ever be truly successful.

Figure 17.3 Process technicians are a major part of a company's quality effort.

CREDIT: thirawatana phaisalratana/ Shutterstock.

Each of the recognized quality gurus—Deming, Juran, Crosby, Peters, and others—offered a different twist on how quality should be approached. By learning from each of them, you can implement a quality program that is above reproach.

Deming taught the importance of applied statistics and the need to constantly improve. In his famous 14 points, he outlined many important leadership tenets that most companies would do well to revisit.

Juran described the importance of working in teams, even touting that all improvements come by way of teamwork and in no other way.

Crosby showed the importance of having a quality attitude, an attitude of zero defects, accepting nothing short of perfection. We all know that being perfect is impossible, but by setting goals at zero defects, continuous improvement can occur.

Peters taught the importance of championing quality and being passionate about quality and excellence. Having the right attitude about quality before you start means you are halfway there when you take your first step.

Wrapping up this introductory chapter, you learned there are a myriad of quality tools to be employed and many ways to employ them. The goal of this textbook is to introduce you to the most commonly employed tools and help familiarize you with these tools so that when you see them again outside the classroom, you will be prepared to take your place in industry as a value-added member of the quality team.

Total Quality Management and Economics

In the business world, the bottom line refers to the profit line on a balance sheet. In Chapter 2, money was thus discussed from the standpoint of the quality process (Figure 17.4).

Figure 17.4 The language of management is money.

CREDIT: Rawpixel.com/Shutterstock.

The four types of quality costs and the negative impact that failures can have on the profitability of the company were covered. The chapter also talked about how describing improvement opportunities in the language of managers helps to accomplish those opportunities.

The cost of poor quality is sometimes called the price of nonconformance. The quality process needs to be focused on improving quality performance in order to reduce waste, reduce costs, keep costs down and profits up, and keep customers happy. Let us face it—we are in it for the money.

The language of money is also the language of economics, and to be an effective employee, you need a basic understanding of how economics works. This chapter provides information on basic economic terms and concepts.

Customer Service

In Chapter 3, you learned about a somewhat softer subject—customers. Too often process technicians are not included in the customer service loop, even though they are the individuals with hands-on experience with the product. Keeping the customer happy is paramount to the success of the business. Unhappy customers find another place to shop. Satisfied customers return for more (Figure 17.5).

It is common for customers to conduct audits and surveys of their suppliers, often on site at your location. When this happens, you as a process technician become a market-facing employee. What you say, how you say it, and how you portray yourself and the company will have a direct impact on that customer's perception of your company. Prepare yourself for this important part of your job. It will not happen every day, but it will happen.

Quality Management Systems—International Standard (ISO)

In Chapter 4, an overview of the ISO 9001 Quality Management System standard was provided. The ISO document describes the basic quality systems that need to function well for a company to manage quality well (Figure 17.6). Some of the elements of this standard—including Identification and Traceability, Control of Monitoring and Measuring Equipment, and Control of Documents—are directly applicable to the role of the process technician.

Figure 17.6 Quality management system umbrella.

You will be expected to follow the policies and procedures of your company in order to comply with this standard. If your company is ISO 9001 registered, and many process industries companies are, then you will be an active participant in the process. When the internal and external auditors come around, you will be audited. The auditors will ask you to explain what you do. Knowing what they are looking for will give you the edge in answering their questions. You may even be given the opportunity to serve as an auditor. If so, take the chance and go for it.

Many quality tools are used to accomplish improvements. The control plans documented by these improvement teams often require the improvements to be integrated back into the daily operating instructions (standard operating procedures) in the plant. These documents form the foundation of the ISO 9001 quality management system at the plant—which is where the process technicians get their instructions for how to run the plant. The ISO 9001 documentation system in your plant is the umbrella under which many pieces of the quality puzzle reside.

Process technicians also encounter ISO 14001 in their work locations. These ISO guidelines, which focus primarily on environmental concerns, are reviewed and compared to ISO 9001 in this chapter.

Quality Management—Quality Reliability Planning

An often-heard complaint about the quality process is that it is too often focused on problems after the fact. In Chapter 5, the topic of quality reliability planning was presented. This topic is a consolidation of three tools used commonly throughout the process industries under many names.

This process is proactive. It establishes a quality design plan before you ever make the first batch of product. In the quality design plan, you work to document the needs of the customers and your plans to measure your own product to meet the customers' needs. The next phase of this process is the quality control plan, in which you document how to control your process in order to achieve the quality necessary to meet the customers' needs. Finally, your process includes a failure mode and effects analysis (FMEA), which is a "what if" kind of tool. In the FMEA, you analyze the probability of failures, the impact of each failure, and your ability to detect each failure before the customer does (Figure 17.7).

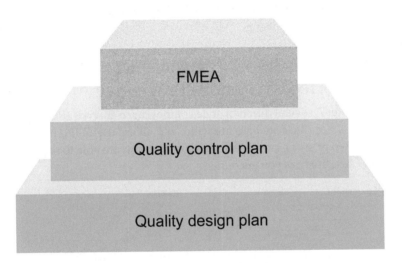

Figure 17.7 Failure mode and effects analyses are based on a quality control plan and quality design plan.

By analyzing your process before something goes wrong, you can implement preventive actions. Although process technicians are not the most common candidates to lead an effort like this, they are valued participants in these types of efforts due to their hands-on knowledge of the process. Who better to help the technical staff understand the likelihood of a particular failure than the person on the shop floor?

Team Skills

As you travel along your quality journey, you rarely travel alone. Much of the work of quality is accomplished in a team environment (Figure 17.8). Quality circles are employed at the control room level to get the various shifts working together on quality issues. Root cause analyses are generally conducted by a multifunctional team.

In Chapter 6, teams in general and several facets of working on teams were discussed. All teams go through growing pains. If you recognize the various stages of team building (forming, storming, norming, and performing), you can work through the stages and get the work done with a minimum of pain.

Several quality tools are available to help teams maintain their focus and function effectively. This chapter discussed the use of a charter, a meeting agenda, an action item log, Gantt charts, and RACI charts to get your teams focused on the task at hand and track your progress along the way. It outlined some basic rules of good team etiquette that most of us

could benefit from practicing on and off the job. Each team member has a role to play. If all team members know their roles and play them well, they can all succeed together.

To be a team player requires you to bring your best self to the workplace. This chapter discusses in depth the qualities and characteristics that will make you valuable to your employer.

Continuous Improvements—Root Cause Analysis (RCA) and Corrective Action/Preventive Action (CPA)

In Chapter 7, several different styles of root cause analysis (RCA) were presented. All these styles of root cause analysis are valid, and all are useful (Figure 17.9). Different companies use different styles, and it is important that you are able to support any type or hybrid of root cause analysis you encounter.

Figure 17.9 Root cause analysis using a fishbone diagram.

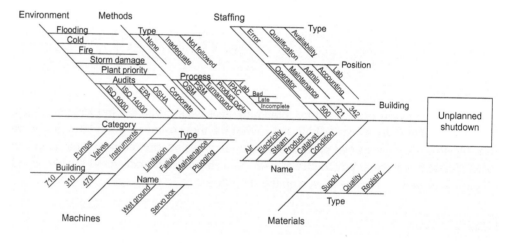

It is common to have process technicians participate on root cause investigation teams because they are the closest to the process. The three styles of root cause analysis most used are the Kepner-Tregoe, Five Whys, and Apollo methodologies. This chapter also briefly presented a couple of other RCA tools that you may see used.

Continuous Improvement—Six Sigma

Chapter 8 described Six Sigma, the tool that may well be the most used approach to process improvement in the process industry today. Six Sigma pulls together the vast majority of the various quality tools into a single improvement strategy (Figure 17.10). Six Sigma improvements are accomplished by using improvement teams.

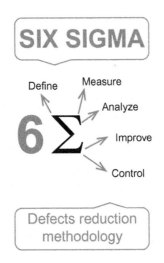

Figure 17.10 Stages of Six Sigma.

CREDIT: Courtesy of Willie L. Myles.

These teams measure the extent of the problem to determine the value proposition. They then analyze the data using control charts, histograms, Pareto charts, and other tools to determine the root cause of the issue. These teams employ designed experiments and other techniques to improve upon the existing process and then develop control plans to ensure that these improvements last.

In many process industry plants, process technicians serve on Six Sigma teams, and some achieve certifications as green belts. Six Sigma is not an improvement tool; it is a formal, disciplined approach to continuous improvement that employs all of the quality tools in your toolbox. It is not a tool itself but an improvement strategy that can be applied to manufacturing processes as well as functional processes such as purchasing or shipping and logistics.

Continuous Improvement—Lean

Your collection of quality tools was rounded out in Chapter 9 with a set of improvement methodologies collectively referred to as Lean. Most of these tools are applicable at the process technician level. Although the traditional implementation of Lean is infused with Japanese buzzwords, this chapter concentrated on the more practical aspects of reducing cycle time, implementing the 5 Ss (a semiformal system for straightening up and organizing the workplace), and driving incremental continuous improvement into the workplace.

You may recall that some of these tools were as simple as making sure there is a place for everything and everything is in its place (Figure 17.11). Since the Lean concepts have grown in popularity in recent years, you can be sure that that you will see these things again.

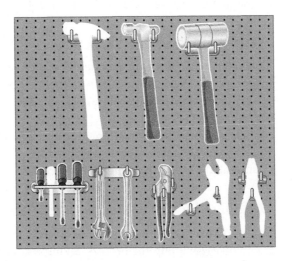

Figure 17.11 Tools organized so each has a space.

Group Problem Solving—Designed Experiments

In Chapter 10, we jumped into the concept of modeling a process and establishing a mathematical link between the process parameters that you physically control and the product that you produce. You learned that sometimes the link is not a simple one-to-one linkage. At times, the impact process parameters have on final product quality is complex because one process variable affects another process variable. This chapter discussed the difference between correlation (a mathematical link) and causation (a true cause/effect relationship) between the inputs and outputs.

Designed experiments are useful statistical tools intended to gather minimum amounts of data to provide maximum amounts of information (Figure 17.12). Designed experiments are not typically set up by process technicians but are almost always carried out by them. How well you follow the design parameters and document what is going on in and around the process has a direct impact on the usefulness of the results obtained. Just so you understand the value of designed experiments, you learned a little bit about how they are set up and how to interpret the results.

Figure 17.12 Catapult.

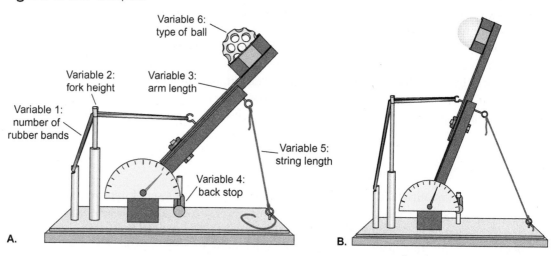

Despite your best efforts, sometimes things just go wrong. When that happens, you have to accept this as an opportunity to improve your process by determining the true root cause. Only by fixing the root cause can you prevent a problem from happening again. Too often companies address the symptoms but never get to the root cause. When this happens, you can be sure the problem will occur again and again and again.

Other Basic Quality Tools

The first few chapters form the foundation upon which much of the house of quality is built. In Chapter 11, the basic toolbox was rounded out with a collection of commonly applied quality tools such as the fishbone diagram, flowcharts, and the Pareto chart (Figure 17.13), just to name a few.

Some of these tools are meant to be applied independently, whereas others are used in a variety of quality improvement activities such as are outlined in other chapters. Each of them is a useful tool when applied in the appropriate way to the appropriate problem.

Data Collection and Representative Sampling

Chapter 12 addresses the importance of true data. An integral part of a process technician's job involves sampling the process, taking readings, and recording data. On a daily basis,

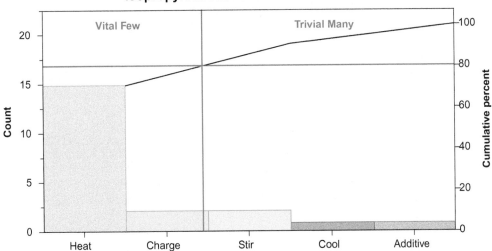

Figure 17.13 Pareto chart.

major operational decisions are made based upon data collected by the process technician. There is an old industry truism "bad data are worse than no data." Therefore, good data collection technique is important to ensure quality data (Figure 17.14). Application of the characteristics of a good data collector, understanding how to ensure a representative sample is taken, and following established sampling procedures are methods that will result in reliable quality data on which to make process improvements.

Figure 17.14 Process technicians take samples.

CREDIT: Photographic Services, Shell International Limited.

Variance and Operating Consistency

In Chapter 13, you took your first steps, laying the groundwork for your knowledge of some of the most common forms of control charts. To understand control charts, you have to understand variation in your process and how to measure it. In this chapter, we discussed in some detail how processes vary and why you should care.

This chapter covered how to use the average (or mean) to measure the center of a process and the standard deviation (or sigma) to measure the spread of the process (Figure 17.15). It introduced the histogram as your first graphical tool to display the variation in your processes and began referring to this picture of your process as a distribution of data. You discovered that there are many types and shapes of distributions depending on the type of process that you have in place.

When your process is behaving normally, exhibiting only normal variation, you deem your process to be stable or in control. When your process is not behaving normally but is

Figure 17.15 Standard deviation graph.

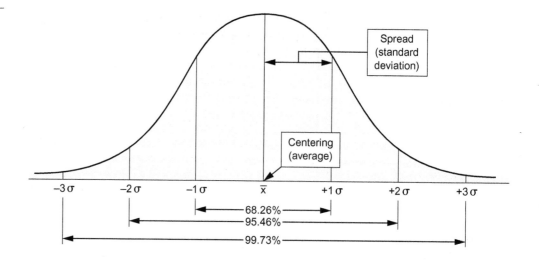

exhibiting some kind of "extra" or abnormal variability, you deem your process to be unstable or out of control. Factors causing this abnormal behavior are referred to as *special causes*.

Variables Control Charts and Interpretation

In Chapter 14, we started building upon the concepts of variation with the variables control charts. These types of control charts are extremely common throughout the process industries (Figure 17.16). You should expect to apply them in your jobs.

The two main control charts focused on in this category were the $\overline{X}R$ control charts and the individuals control charts, both of which are built upon the normal distribution of data. This chapter also introduced some of the other variables control charts that you may see in practice, such as zone charts and exponentially weighted moving average (EWMA) charts, although the construction of these last two types was left for others to explain due to their complexity.

Figure 17.16 Run chart showing upper and lower control limits.

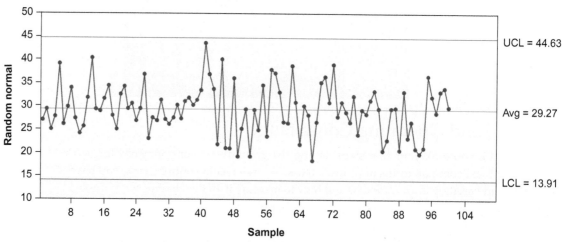

Attributes Control Charts and Interpretation

In Chapter 15, you continued the study of control charts by adding control charts from nonnormal distributions such as the C and U charts (Figure 17.17), which are built on the Poisson distribution, and the P and NP charts, which are built on the binomial distribution.

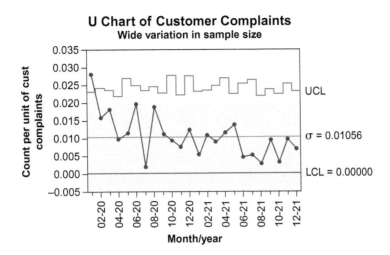

Figure 17.17 Computer-generated U chart of customer complaints.

As you recall, the Poisson distribution is the distribution of counting data and is useful for counting of defects and is (or should be) employed extensively in the charting of safety and injury statistics. The binomial distribution is the distribution of defectives and is most commonly used when charting things with only two possible outcomes, such as pass/fail data.

Wrapping up this series of chapters on variation and how to measure, plot, and interpret that variation, you learned that regardless of what type of control chart you employ, the interpretation rules are the same.

The three main indicators are: (1) points outside the control limits, (2) shifts up or down in the data, and (3) trends up or down within the limits. These three indicators suggest that something abnormal is going on in the process, which gives you the opportunity to investigate and learn from the occurrence, ultimately improving your process.

Process Capability

In Chapter 16, the concept of process variability was carried to the next level by introducing several process capability indices such as Cpk, Cp, and Ppk. These provide a measure of relative goodness of a process compared with the specification limits established with customers (Figure 17.18).

Figure 17.18 Process capability guidelines.

Process capability indexes rules of thumb

<1.0 - Opportunity for improvement
1.0–1.33 - Okay but borderline
>1.33 - Good - this is what we want

If process variation is much less than what the customer desires, then the process is good. If process variation is more than what the customer desires, there is an opportunity for improvement.

In general, we can grade process variation using these indices: Any index below 1 is an opportunity for improvement, an index between 1 and 1.33 is borderline, and an index above 1.33 is good. You learned about the natural variation that occurs in measurement processes as well.

Summary

This book has provided an overview of quality-related topics, many of which are the subject of entire books themselves. It is intended to provide insights that will allow you to become a top-quality employee and to help create a top-quality environment.

CREDIT: Robert Adrian Hillman/Shutterstock.

What you should walk away with is this:

1. Quality is an important part of your role as a process technician. The data you record and submit will be used to steer decision making.

2. There are many pieces to the quality puzzle, and these pieces all fit together to help you manage and improve your processes. Employee involvement and teamwork are crucial to successful operations.

3. There are tools available in your quality toolbox for many different tasks. Applying the right tool to the task is part of your job.

4. This is not the end of your quality journey; it is just the beginning. Quality work leads to quality product and personal satisfaction. Go out and improve yourself and your process every day.

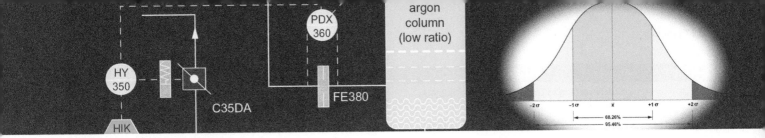

Appendix

Answers to Checking Your Knowledge Questions

Chapter 1	Answer	LO #
1.	See Key Terms list.	1.4
2.	False	1.1
3.	d. All of the above	1.4
4.	True	1.1
5.	c. Crosby, Deming, and Juran	1.3
6.	b. Training, teams, statistics, management systems, and management commitment	1.4
7.	True	1.2
8.	I. Control charts — D II. Process capability — A III. Pareto charts — C IV. Designed experiments — F V. Management systems — B VI. Root cause analysis — E	1.4
9.	False	1.2
10.	b. The evaluation of data to determine the root cause of a problem	1.3, 1.4
11.	a. The process of selecting the proper fix (remedy) for a problem	1.3, 1.4
12.	True	1.3, 1.4
13.	c. Crosby	1.3
14.	False	1.3
15.	False	1.4

Chapter 2		
1.	See Key Terms list.	2.1, 2.2, 2.3
2.	c. Money	2.1
3.	d. Describe the value of the project to managers in terms they understand	2.1
4.	a. Cost of capital	2.2
5.	d. All of the above	2.4
6.	d.	2.2
7.	c. Costs of quality improvement projects	2.2
8.	a. The portion of the plant that makes poor-quality product	2.3
9.	False	2.1
10.	True	2.2
11.	False	2.1
12.	False	2.2
13.	False	2.1

14.	True	2.3
15.	True	2.2
16.	a. Skills b. Talent c. Morale e. Initiative	2.4
17.	False	2.4

Chapter 3		
1.	See Key Terms list.	3.4
2.	d. All of these plus more	3.5
3.	b. SPC	3.2
4.	c. Customer	3.2
5.	a. Internal customer	3.3
6.	c. External customer	3.3
7.	d. Schedule high-impact meetings during the customer visit.	3.5
8.	d. 100 percent is always your target.	3.2
9.	False	3.1
10.	True	3.2
11.	True	3.1, 3.4
12.	True	3.3
13.	True	3.5
14.	True	3.5
15.	False	3.1

Chapter 4		
1.	See Key Terms list.	4.1, 4.2, 4.5
2.	True	4.3
3.	a. Competence, training, and awareness d. Identification and traceability e. Monitoring and measurement of product f. Monitoring and measurement of processes	4.2
4.	b. Environmental management	4.3
5.	b. An audit.	4.4
6.	b. Ensuring the management system meets the requirements of the standard c. Ensuring the management system is effectively implemented	4.4
7.	a. Medical records	4.1, 4.2, 4.4

Chapter 4	Answer	LO #
8.	d. It is never okay to provide false information.	4.2, 4.4
9.	b. Treat the auditor with respect	4.2, 4.4
10.	d. Looking around for things	4.2, 4.4
11.	a. Pointing the blame at someone else	4.2, 4.4
12.	True	4.2, 4.4
13.	a. Malcolm Baldrige National Quality Award	4.5
14.	False	4.3
15.	True	4.2
16.	True	4.5

Chapter 5		
1.	See Key Terms list.	5.1, 5.2, 5.3, 5.4
2.	c. Quality reliability planning.	5.1
3.	b. Quality control plan.	5.1
4.	a. Failure mode and effects analysis.	5.4
5.	d. Quality design plan.	5.1
6.	d. All of the above	5.1
7.	True	5.3
8.	d. Quality design plan (QDP)	5.2
9.	True	5.2
10.	b. Quality control plan (QCP)	5.2
11.	a. Failure mode and effects analysis.	5.4
12.	False	5.4

Chapter 6		
1.	See Key Terms list.	6.1
2.	True	6.1
3.	a. Patience b. Flexibility d. Confidence e. Adaptability	6.1
4.	b. Highly trained participants	6.1
5.	d. All of these and more	6.2
6.	a. forming.	6.3
7.	b. norming.	6.3
8.	b. storming.	6.3
9.	e. Any of the above	6.2, 6.4
10.	True	6.5
11.	True	6.5
12.	False	6.1
13.	True	6.1
14.	True	6.5, 6.7
15.	False	6.6, 6.7
16.	True	6.6, 6.7
17.	False	6.6

Chapter 7		
1.	See Key Terms list.	7.1
2.	c. Kepner-Tregoe	7.3

Chapter 7		
3.	a. Five Whys	7.3
4.	b. Apollo	7.3
5.	d. All of the above	7.3
6.	c. Current reality tree	7.3, 7.4
7.	d. Interrelationship digraph	7.3, 7.4
8.	b. Cause-and-effect diagram	7.3, 7.4
9.	False	7.2
10.	True	7.1
11.	False	7.1
12.	True	7.2
13.	True	7.1, 7.2
14.	True	7.1, 7.2
15.	True	7.1, 7.2, 7.4

Chapter 8		
1.	See Key Terms list.	8.1, 8.2
2.	False	8.1
3.	a. Shewhart c. Deming	8.1
4.	a. green belts.	8.2
5.	b. black belts.	8.2
6.	c. master black belts.	8.2
7.	b. a formal, disciplined, top-driven approach to continuous improvement.	8.3
8.	d. 3.4 defects per million opportunities.	8.2
9.	False	8.4
10.	True	8.2
11.	c. Design	8.3
12.	b. Analyze	8.3
13.	c. Improve	8.3
14.	d. Control	8.3
15.	False	8.4
16.	True	8.4

Chapter 9		
1.	See Key Terms list.	9.2
2.	a. a philosophy of working without waste.	9.1
3.	c. Toyota	9.1
4.	d. Overtime	9.1
5.	a. Transport c. Overproduction d. Motion e. Excess processing	9.1
6.	b. value stream maps.	9.2
7.	c. the total process time divided by the number of customer orders per block of time.	9.2
8.	a. Customer willing to pay for the service c. Transforms the product d. Is free from defects	9.2

Chapter 9	Answer	LO #
9.	False	9.1
10.	True	9.2
11.	True	9.2
12.	True	9.1, 9.2

Chapter 10		
1.	See Key Terms list.	10.3
2.	True	10.1
3.	a. The range of data that was used to calculate the model	10.1
4.	c. $Y = f(x)$	10.2
5.	b. This is a negative correlation.	10.2
6.	a. This is a positive correlation.	10.2
7.	b. This is a strong correlation.	10.2
8.	c. This is a zero correlation.	10.2
9.	a. This is a weak correlation.	10.2
10.	d. 0.75	10.2
11.	False	10.3
12.	d. 4	10.4
13.	b. the percentage of the variability in the data explained by the model.	10.3
14.	True	10.2
15.	False	10.2
16.	True	10.5
17.	False	10.4
18.	False	10.2
19.	True	10.5

Chapter 11		
1.	See Key Terms list.	11.1, 11.2, 11.3, 11.4
2.	b. Quantitative	11.1
3.	False	11.1
4.	a. Process steps	11.4
5.	a. the 80/20 rule.	11.4
6.	b. brainstorming.	11.2
7.	c. Fishbone diagram	11.3
8.	d. Motors	11.3
9.	b. the vital few.	11.4
10.	a. the trivial many.	11.4
11.	c. Trend charts	11.4
12.	a. Frequency	11.4
13.	b. Cumulative percentage	11.4
14.	False	11.4
15.	False	11.4
16.	True	11.2

Chapter 12		
1.	See Key Terms list.	
2.	True	12.1
3.	b. Particle size	12.1

Chapter 12		
4.	False	12.1
5.	c. Analytic study	12.2
6.	False	12.2
7.	d. Enumerative study	12.2
8.	d. Observing	12.3
9.	a. Source of the data c. How the data will be measured d. Method for collecting	12.3
10.	True	12.4
11.	d. Irreconcilable data	12.4
12.	b. Recirculation	12.5
13.	b. Level of accuracy	12.5
14.	False	12.5
15.	False	12.6
16.	a. Paying attention to detail c. Using an analytical mindset d. Accurately documenting the data collected	12.6
17.	True	12.5

Chapter 13		
1.	See Key Terms list.	13.1, 13.3, 13.4
2.	True	13.1
3.	c. display the distribution of data in a bar chart format.	13.1
4.	e. all of the above	13.1
5.	True	13.1
6.	a. Standardized procedures b. Training e. Application of statistical tools to recognize special cause variation	13.1
7.	False	13.1
8.	a. Mean	13.2
9.	b. Standard deviation	13.3
10.	c. σ	13.3
11.	a. Σ	13.3
12.	b. 68 percent	13.3
13.	c. 95 percent	13.5
14.	d. 99.7 percent	13.5
15.	c. Poisson	13.5
16.	c. Binomial	13.5
17.	True	13.4
18.	False	13.4

Chapter 14		
1.	See Key Terms list.	14.4
2.	a. They help distinguish between common cause and special cause variation.	14.1
3.	False	14.1
4.	c. time	14.1

Chapter 14	Answer	LO #
5.	b. Individuals control charts use individual measurements, whereas $\overline{X}R$ charts use averages of measurements divided into subgroups.	14.4
6.	b. when your data are grouped into rational subgroups.	14.4
7.	c. when your data cannot be grouped into rational subgroups.	14.4
8.	a. They make setting up, plotting, monitoring, and interpretation of control charts as simple as possible.	14.5
9.	a. EWMA	14.5
10.	e. Whichever one is the best fit for the particular need	14.1, 14.4, 14.5
11.	c. normal	14.1, 14.4, 14.5
12.	True	14.1

Chapter 15		
1.	See Key Terms list.	15.1, 15.2, 15.3
2.	b. Counts of customer returns	15.2
3.	c. Proportions of defective drums	15.2
4.	a. C	15.2
5.	b. U	15.2
6.	d. P	15.2
7.	c. NP	15.2
8.	b. 10 percent	15.2
9.	a. Out of control c. Shifts d. Trends	15.3

10.	b. constant sample size allows you to have only one set of control limits.	15.2
11.	I. Out of control—b II. Shifts—c III. Trends—a	15.3
12.	True	15.1, 15.2
13.	False	15.1, 15.2
14.	True	15.1, 15.2
15.	False	15.1, 15.2

Chapter 16		
1.	See Key Terms list.	16.1, 16.2, 16.3, 16.5
2.	c. 1.0	16.2
3.	a. 0.67	16.2
4.	a. 0.67	16.2
5.	c. 5.0	16.4
6.	c. 5.0	16.4
7.	False	16.2
8.	d. Changing your PPE	16.3
9.	c. Greater than 1.0	16.1, 16.2
10.	False	16.4
11.	b. The variability increases.	16.6
12.	False	16.4
13.	c. 1.33	16.1, 16.2
14.	False	16.5
15.	c. Analyze data	16.4
16.	d. 40 percent	16.6

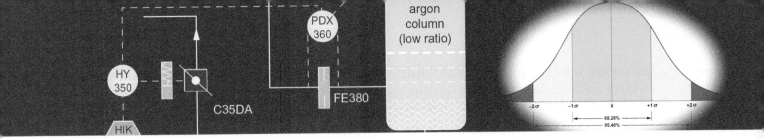

Glossary

σ lowercase Greek letter sigma; used in mathematics to denote the standard deviation.

Σ uppercase Greek letter sigma; used in mathematics to indicate a summation.

5 Ss a floor-level improvement methodology taken from five Japanese words: Seiri, Seiton, Seiso, Seiketsu, and Shitsuke.

Accreditation the formal recognition by an independent body, generally known as an accreditation body, that a prospective organization operates according to international standards.

Accuracy exactness; the quality of being exact.

Analytic study a statistical study aimed at prediction of future results or effects, that begins with evaluation of the results or effects of a cause-system and ends with probable inferences (implicit predictions) about other results or effects not yet observed.

Appraisal costs the cost of checking performance to attempt to catch mistakes before they get out the door.

Attributes data data from a discrete distribution in which only whole numbers (counts) are possible.

Audit a review of the management system to see whether the management system is effectively implemented.

Auditee an entity or person being audited.

Auditor a person conducting an audit.

Binomial distribution distribution of discrete data that are made up of data that have only two choices (e.g., pass/fail).

Black belt team leader on a Six Sigma team who has training in advanced quality improvement tools, plus training in project management.

Brainstorming a group exercise designed to solicit a large number of ideas in a short amount of time.

C chart control chart for counts of defects (Poisson distribution) with a constant sample size.

Central tendency center or middle of a distribution of data; it is measured as the mean, median, or mode.

Champion in Six Sigma terms, a manager responsible for the overall effectiveness of the Six Sigma implementation at a business or functional level.

Common cause variation variables in a process that cause the process to vary and are built in and inherent to the process. The variation of the process due to common causes is the variation that is always there when things are running normally.

Confidence level the level of probability that a sample would represent the population.

Continuous data numbers that can be of any value or fraction of a value.

Continuous improvement an ongoing effort to improve products, services, or processes using critical thinking skills.

Control chart graphical presentation of statistical data that can be used to follow and identify process variations and problems.

Correlation the mutual relationship of two or more things; the degree to which two or more attributes tend to vary together.

Design for Six Sigma (DFSS) the application of appropriate Six Sigma tools to new process designs rather than to the improvement of existing processes.

Discrete data numbers that are not continuous (e.g., whole numbers). Also known as *variables data*.

Documentation written or electronic information that records actions or defines how a task is to be accomplished.

Enumerative study a statistical study aimed at the description of a well-defined population; it is an investigation that begins with one or more random samples from a population and ends with inferences about the properties of other elements that were not (but could have been) sampled.

Expectations desires of the customer that might not prevent the product from working as intended but would cause dissatisfaction if not met.

Exponentially weighted moving average (EWMA) chart a special type of control chart that takes previous data into account when evaluating each point; also capable of detecting small shifts in the mean.

External failure costs the cost of mistakes that make it out the door to customers.

Failure mode and effects analysis (FMEA) the third and final step in the quality reliability planning process, in which you examine every possible failure mode for the process in order to understand the consequences of process failures from the customers' perspective.

Fishbone diagram a graphic presentation device whereby ideas (or causes) are grouped into common categories such as manpower, methods, machines, and materials. Also known as an *Ishikawa* or *cause-and-effect diagram*.

Flowcharts documents that represent a sequence of operations schematically or visually.

Green belt team member on a Six Sigma team that has training in the basic quality improvement tools.

Hidden factory the portion of the factory that produces non-value-added product.

Histogram a type of graphical presentation of data.

Internal failure costs the cost of mistakes that were caught before they got out the door.

International Organization for Standardization (ISO) federation made up of over 160 member countries that have agreed to publish common standard methodologies for various aspects of conducting global business.

ISO 14001 the internationally recognized environmental management system standard.

ISO 9001 the internationally recognized quality management system standard.

ISO registered the recognition granted to a company by an accredited registrar when the registrar has verified that the company has effectively implemented the management system.

Just-in-time (JIT) process by which the company creates a product for the customer when the customer needs it.

Kaizen Japanese term for continuous improvement or incremental improvement.

Kanban a signal card used to pull production.

Lower control limit (LCL) bottom limit in quality control for data points below the control (average) line in a control chart.

Lower specification limit (LSL) the lowest level allowed by the customer for a specific quality measure.

Malcolm Baldrige National Quality Award (MBNQA) a U.S.–based award established by Congress in 1987 that recognizes not only the effectiveness of a management system but also the company results achieved as a result of its management system.

Management system a collection of activities, usually but not always documented, that are intended to ensure that products and services meet specified needs.

Margin of error the number of times (in +/−%) a sample would represent the whole, given a specific confidence level.

Master black belt Six Sigma expert who has training in many advanced quality improvement tools and is responsible for coaching black belts and providing training to green belts and black belts.

Mean mathematical average of a set of data.

Measurement capability index (Cm) the ratio of your specification band to your measurement process variability.

Median geometric center of a data set.

Mode number that occurs most frequently in a data set.

Moving range (MR) difference between two consecutive points.

Muda Japanese term for waste.

Mura Japanese term for unevenness.

Muri Japanese term for overburden.

Nonconformance evidence of deviation from documented requirements.

Normal distribution distribution of continuous data that is bell shaped. The data in a normal distribution are distributed such that approximately 68 percent of the data fall within 1 standard deviation of the mean, approximately 95 percent of the data fall within 2 standard deviations of the mean, and approximately 99.7 percent of the data fall within 3 standard deviations of the mean. Also known as a *Gaussian distribution*.

NP chart control chart for counts of defectives (binomial distribution) with a constant sample size.

Operational definition a description that defines or gives meaning to a variable by describing how it will be measured.

Out of control points outside the three sigma (σ) control limits.

P chart control chart for counts of defectives (binomial distribution) with a variable sample size.

Pareto charts graphics that rank causes from most to least significant; they represent the 80-20 rule, which says that most undesired effects come from relatively few causes.

Performance model the mathematical model that predicts how your process will perform based on the input variables.

Plan-do-check-act (PDCA) cycle a cycle for continuous improvement that implements the following steps: planning, doing, checking, and acting. Some companies use *plan-do-study-act* or a version called *standardize-do-check-act (SDCA)*.

Poisson distribution distribution of discrete data that consist of counting data. In this case, only whole numbers are present.

Precision a measurement of how closely the sample results can be duplicated.

Prevention costs the costs of improving processes so that they do not have to be checked to catch failures because everything was done right the first time.

Price of nonconformance (PONC) a term coined by Philip Crosby used to describe the costs of poor performance and failing to meet the customers' needs, also called *cost of poor quality (COPQ)*.

Process capability index (Cpk) the ratio of the customer's specification range to your process variability taking the average (process center) into account; this index assumes the data have been "cleansed" of special causes.

Process performance index (Ppk) the same calculation as the Cpk but without the assumption that the data have been cleansed of special causes.

Process potential C_{pl} process potential for processes that have only a lower specification limit.

Process potential C_{pu} process potential for processes that have only an upper specification limit.

Process potential index (Cp) the simple ratio of the customer's specification range to normal process variability without taking the average into account.

Process Safety Management (PSM) an OSHA standard, 29CFR1910.119, concerning process safety management of highly hazardous chemicals.

Quality conformance to all aspects of customer requirements.

Quality circle a group of people, typically organized by logical structure within a company, who work together for a common goal.

Quality control plan (QCP) the second step in the quality reliability planning process. Process control parameters are specified to ensure the process is capable of meeting the needs of the customer. Process capability is captured at this stage.

Quality design plan (QDP) the first step in the quality reliability planning process. The needs of the customer are translated into specific requirements (specifications), and the measurement techniques are specified to ensure the needs are met.

Quality reliability planning (QRP) a procedure designed to ensure that when a new product is being introduced, proper communications occur between all of the functional groups and that a system is in place that allows the product to meet the customers' expectations consistently. The procedure consists of the quality design plan, the quality control plan, and the failure mode and effects analysis.

Range (R) the limits, from minimum point to maximum, for any operating condition or specification. Usually used to refer to temperature, pressure, level, specification, flow, or product purity.

Records evidence that a task has been accomplished.

Regression the statistical analysis that generates a mathematical representation of the correlation between two or more variables.

Representative sample a subset of a population that seeks to accurately reflect the characteristics of the larger group.

Requirements the criteria that describe the traits that a product or service must possess in order to function as intended. Performance characteristics of your product.

Risk priority number (RPN) a calculated number within the failure mode and effects analysis process that helps prioritize the actions that need to be taken to minimize the risks of process failures and consequences to the customer.

Root cause analysis (RCA) the analysis of data to determine the true root cause of a problem which, when identified and corrected, eliminates a problem forever. Also called *root cause investigation*.

Sample a representative portion of a material collected for analysis.

Shifts runs of eight points above or below the average.

Size of population the number of units which make up the whole.

Special cause variation event-related items in the process that cause the process to vary outside of its normal operation. These causes are in addition to the common cause variation sources. Also known as *assignable cause variation*.

Specifications a subset of the requirements that the customer puts into the transaction contract. May also be the translation of the requirement into a measurable output for the purpose of tracking compliance.

Spread the broadness of the distribution of the data set; it is measured as the difference between the highest and lowest values in the data set or by using the standard deviation.

Stable process   state of a process when only common cause variation is present; also stated as the absence of special or assignable causes. A stable process is, by definition, predictable.

Standard deviation   statistical measure of the spread of a set of data. Represented by lower case *sigma* (σ).

Statistical process control (SPC)   statistical procedures that keep track of a process in order to reduce variation and improve quality.

Statistical quality control (SQC)   the application of statistical techniques to the output (quality) of a process.

Synergy   the phenomenon in which the total effect of a whole is greater than the sum of its individual parts.

Takt time   a German word that refers to keeping a beat. In quality discussions, it is the available production time divided by the customer demand.

Team   a small group of people, with complementary skills, committed to a common set of goals and tasks.

Total quality management (TQM)   the pulling together of many different components of quality into a single program (sometimes referred to as TQC—total quality control).

Trend charts   graphic representations of a direction in a set of statistical data at a particular time. Also, a computer program which allows process variables to be monitored for a set period of time.

Trends   the direction in a set of statistical data at a particular time; runs of eight points going up or going down.

U chart   control chart for counts of defects (Poisson distribution) with a variable sample size.

Unstable process   state of your process when both common cause and special cause variation are present. An unstable process is, by definition, unpredictable.

Upper control limit (UCL)   the highest level of acceptable quality for a product or service.

Upper specification limit (USL)   the highest level allowed by the customer for a specific quality measure.

Value stream map (VSM)   a high-level flowchart of a process. Included within each step of the process is the amount of time and the value it adds from the customer's perspective.

Variables data   data from a distribution in which any number or fraction of a number is possible.

x-bar and R ($\overline{X}R$)   a control chart that displays both the mean value (X) as well as the range (R) to indicate changes in the mean value and dispersion over an established period of time.

x-bar   used in mathematics to indicate the mean.

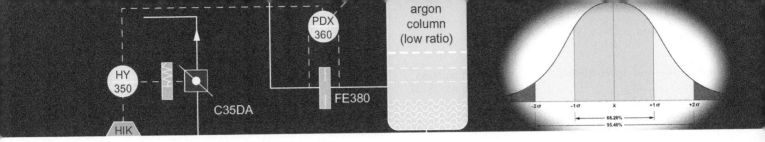

Index